Lecture Notes in Mathematics　　　2061

Editors:
J.-M. Morel, Cachan
B. Teissier, Paris

For further volumes:
http://www.springer.com/series/304

Lévy in Stanford (Permission granted by G.L. Alexanderson)

"Lévy Matters" is a subseries of the Springer Lecture Notes in Mathematics, devoted to the dissemination of important developments in the area of Stochastics that are rooted in the theory of Lévy processes. Each volume will contain state-of-the-art theoretical results as well as applications of this rapidly evolving field, with special emphasis on the case of discontinuous paths. Contributions to this series by leading experts will present or survey new and exciting areas of recent theoretical developments, or will focus on some of the more promising applications in related fields. In this way each volume will constitute a reference text that will serve PhD students, postdoctoral researchers and seasoned researchers alike.

Editors

Ole E. Barndorff-Nielsen
Thiele Centre for Applied Mathematics
   in Natural Science
Department of Mathematical Sciences
Aarhus University
8000 Aarhus C, Denmark
oebn@imf.au.dk

Jean Jacod
Institut de Mathématiques de Jussieu
CNRS-UMR 7586
Université Paris 6 - Pierre et Marie Curie
75252 Paris Cedex 05, France
jean.jacod@upmc.fr

Jean Bertoin
Institut für Mathematik
Universität Zürich
8057 Zürich, Switzerland
jean.bertoin@math.uzh.ch

Claudia Klüppelberg
Zentrum Mathematik
Technische Universität München
85747 Garching bei München, Germany
cklu@ma.tum.de

Managing Editors

Vicky Fasen
Department of Mathematics
ETH Zürich
8092 Zürich, Switzerland
vicky.fasen@math.ethz.ch

Robert Stelzer
Institute of Mathematical Finance
Ulm University
89081 Ulm, Germany
robert.stelzer@uni-ulm.de

The volumes in this subseries are published under the auspices of the Bernoulli Society.

Serge Cohen • Alexey Kuznetsov
Andreas E. Kyprianou • Victor Rivero

Lévy Matters II

Recent Progress in Theory and Applications:
Fractional Lévy Fields, and Scale Functions

 Springer

Serge Cohen
Université Paul Sabatier
Institut Mathématique de Toulouse
Toulouse Cedex 4, France

Andreas E. Kyprianou
University of Bath
Department of Mathematical Sciences
Bath, United Kingdom

Alexey Kuznetsov
York University
Department of Mathematics
and Statistics
Toronto, ON, Canada

Victor Rivero
CIMAT A.C.
Guanajuato, Col. Valenciana, Mexico

ISBN 978-3-642-31406-3   ISBN 978-3-642-31407-0 (eBook)
DOI 10.1007/978-3-642-31407-0
Springer Heidelberg New York Dordrecht London

Lecture Notes in Mathematics ISSN print edition: 0075-8434
                                    ISSN electronic edition: 1617-9692

Library of Congress Control Number: 2012945523

Mathematics Subject Classification (2010): Primary: 60G10, 60G40, 60G51, 60G70, 60J10, 60J45
                                    Secondary: 91B28, 91B70, 91B84

© Springer-Verlag Berlin Heidelberg 2012
This work is subject to copyright. All rights are reserved by the Publisher, whether the whole or part of
the material is concerned, specifically the rights of translation, reprinting, reuse of illustrations, recitation,
broadcasting, reproduction on microfilms or in any other physical way, and transmission or information
storage and retrieval, electronic adaptation, computer software, or by similar or dissimilar methodology
now known or hereafter developed. Exempted from this legal reservation are brief excerpts in connection
with reviews or scholarly analysis or material supplied specifically for the purpose of being entered
and executed on a computer system, for exclusive use by the purchaser of the work. Duplication of
this publication or parts thereof is permitted only under the provisions of the Copyright Law of the
Publisher's location, in its current version, and permission for use must always be obtained from Springer.
Permissions for use may be obtained through RightsLink at the Copyright Clearance Center. Violations
are liable to prosecution under the respective Copyright Law.
The use of general descriptive names, registered names, trademarks, service marks, etc. in this publication
does not imply, even in the absence of a specific statement, that such names are exempt from the relevant
protective laws and regulations and therefore free for general use.
While the advice and information in this book are believed to be true and accurate at the date of
publication, neither the authors nor the editors nor the publisher can accept any legal responsibility for
any errors or omissions that may be made. The publisher makes no warranty, express or implied, with
respect to the material contained herein.

Printed on acid-free paper

Springer is part of Springer Science+Business Media (www.springer.com)

Preface

This second volume of the series "Lévy Matters" consists of two surveys of two topical areas, namely Fractional Lévy Fields by Serge Cohen and the Theory of Scale Functions for Spectrally Negative Lévy Processes by Alexey Kuznetsov, Andreas Kyprianou and Victor Rivero.

Roughly speaking, irregularity is a crucial aspect of random phenomena that appears in many different contexts. An important issue in this direction is to offer tractable mathematical models that encompass the variety of observed behaviours in applications. Fractional Lévy fields are constructed by integration of Lévy random measures; somehow they interpolate between Gaussian and stable random fields. They exhibit a number of interesting features including local asymptotic self-similarity and multi-fractional aspects. Calibration techniques and simulation of fractional Lévy fields constitute important elements for many applications.

A real-valued Lévy process is spectrally negative when it has no positive jumps. In this situation, the distribution of several variables related to the first exit-time from a bounded interval can be specified in terms of the so-called scale functions; the latter also play a fundamental role in other aspects of the theory. Scale functions are characterized by their Laplace transform, but in general no explicit formula is known, and therefore it is crucial in many applications to gather information about their asymptotic behaviour and regularity and to provide efficient numerical methods to compute them.

Aarhus, Denmark   Ole E. Barndorff-Nielsen
Zürich, Switzerland   Jean Bertoin
Paris, France   Jean Jacod
Munich, Germany   Claudia Küppelberg

From the Preface to Lévy Matters I

Over the past 10–15 years, we have seen a revival of general Lévy processes theory as well as a burst of new applications. In the past, Brownian motion or the Poisson process had been considered as appropriate models for most applications. Nowadays, the need for more realistic modelling of irregular behaviour of phenomena in nature and society like jumps, bursts and extremes has led to a renaissance of the theory of general Lévy processes. Theoretical and applied researchers in fields as diverse as quantum theory, statistical physics, meteorology, seismology, statistics, insurance, finance and telecommunication have realized the enormous flexibility of Lévy models in modelling jumps, tails, dependence and sample path behaviour. Lévy processes or Lévy-driven processes feature slow or rapid structural breaks, extremal behaviour, clustering and clumping of points.

Tools and techniques from related but distinct mathematical fields, such as point processes, stochastic integration, probability theory in abstract spaces and differential geometry, have contributed to a better understanding of Lévy jump processes.

As in many other fields, the enormous power of modern computers has also changed the view of Lévy processes. Simulation methods for paths of Lévy processes and realizations of their functionals have been developed. Monte Carlo simulation makes it possible to determine the distribution of functionals of sample paths of Lévy processes to a high level of accuracy.

This development of Lévy processes was accompanied and triggered by a series of Conferences on Lévy Processes: Theory and Applications. The First and Second Conferences were held in Aarhus (1999, 2002), the Third in Paris (2003), the Fourth in Manchester (2005) and the Fifth in Copenhagen (2007).

To show the broad spectrum of these conferences, the following topics are taken from the announcement of the Copenhagen conference:

- Structural results for Lévy processes: distribution and path properties
- Lévy trees, superprocesses and branching theory
- Fractal processes and fractal phenomena
- Stable and infinitely divisible processes and distributions

- Applications in finance, physics, biosciences and telecommunications
- Lévy processes on abstract structures
- Statistical, numerical and simulation aspects of Lévy processes
- Lévy and stable random fields

At the Conference on Lévy Processes: Theory and Applications in Copenhagen the idea was born to start a series of Lecture Notes on Lévy processes to bear witness of the exciting recent advances in the area of Lévy processes and their applications. Its goal is the dissemination of important developments in theory and applications. Each volume will describe state-of-the-art results of this rapidly evolving subject with special emphasis on the non-Brownian world. Leading experts will present new exciting fields, or surveys of recent developments, or focus on some of the most promising applications. Despite its special character, each article is written in an expository style, normally with an extensive bibliography at the end. In this way each article makes an invaluable comprehensive reference text. The intended audience are PhD and postdoctoral students, or researchers, who want to learn about recent advances in the theory of Lévy processes and to get an overview of new applications in different fields.

Now, with the field in full flourish and with future interest definitely increasing it seemed reasonable to start a series of Lecture Notes in this area, whose individual volumes will appear over time under the common name "Lévy Matters," in tune with the developments in the field. "Lévy Matters" appears as a subseries of the Springer Lecture Notes in Mathematics, thus ensuring wide dissemination of the scientific material. The mainly expository articles should reflect the broadness of the area of Lévy processes.

We take the possibility to acknowledge the very positive collaboration with the relevant Springer staff and the editors of the LN series and the (anonymous) referees of the articles.

We hope that the readers of "Lévy Matters" enjoy learning about the high potential of Lévy processes in theory and applications. Researchers with ideas for contributions to further volumes in the Lévy Matters series are invited to contact any of the editors with proposals or suggestions.

Aarhus, Denmark	Ole E. Barndorff-Nielsen
Zürich, Switzerland	Jean Bertoin
Paris, France	Jean Jacod
Munich, Germany	Claudia Küppelberg

Contents

A Short Biography of Paul Lévy

A volume of the series "Lévy Matters" would not be complete without a short sketch about the life and mathematical achievements of the mathematician whose name has been borrowed and used here. This is more a form of tribute to Paul Lévy, who not only invented what we call now Lévy processes, but also is in a sense the founder of the way we are now looking at stochastic processes, with emphasis on the path properties.

Paul Lévy was born in 1886 and lived until 1971. He studied at the Ecole Polytechnique in Paris and was soon appointed as professor of mathematics in the same institution, a position that he held from 1920 to 1959. He started his career as an analyst, with 20 published papers between 1905 (he was then 19 years old) and 1914, and he became interested in probability by chance, so to speak, when asked to give a series of lectures on this topic in 1919 in that same school: this was the starting point of an astounding series of contributions in this field, in parallel with a continuing activity in functional analysis.

Very briefly, one can mention that he is the mathematician who introduced characteristic functions in full generality, proving in particular the characterization theorem and the first "Lévy's theorem" about convergence. This naturally led him to study more deeply the convergence in law with its metric, and also to consider sums of independent variables, a hot topic at the time: Paul Lévy proved a form of the 0-1 law, as well as many other results, for series of independent variables. He also introduced stable and quasi-stable distributions and unravelled their weak and/or strong domains of attractions, simultaneously with Feller.

Then we arrive at the book *Théorie de l'addition des variables aléatoires*, published in 1937, and in which he summaries his findings about what he called "additive processes" (the homogeneous additive processes are now called Lévy processes, but he did not restrict his attention to the homogeneous case). This book contains a host of new ideas and new concepts: the decomposition into the sum of jumps at fixed times and the rest of the process; the Poissonian structure of the jumps for an additive process without fixed times of discontinuities; the "compensation" of those jumps so that one is able to sum up all of them; the fact that the remaining continuous part is Gaussian. As a consequence, he implicitly gave the formula providing the form of all additive processes without fixed discontinuities, now called the Lévy–Itô formula, and he proved the Lévy–Khintchine formula for the characteristic functions of all infinitely divisible distributions. But, as fundamental as all those results are, this book contains more: new methods, like martingales which, although not given a name, are used in a fundamental way; and also a new way of looking at processes, which is the "pathwise" way: he was certainly the first to understand the importance of looking at and describing the paths of a stochastic process, instead of considering that everything is encapsulated into the distribution of the processes.

This is of course not the end of the story. Paul Lévy undertook a very deep analysis of Brownian motion, culminating in his book *Processus stochastiques et mouvement brownien* in 1948, completed by a second edition in 1965. This is a remarkable achievement, in the spirit of path properties, and again it contains so many deep results: the Lévy modulus of continuity, the Hausdorff dimension of the path, the multiple points, and the Lévy characterization theorem. He introduced local time, and proved the arc-sine law. He was also the first to consider genuine stochastic integrals, with the area formula. In this topic again, his ideas have been the origin of a huge amount of subsequent work, which is still going on. It also laid some of the basis for the fine study of Markov processes, like the local time again, or the new concept of instantaneous state. He also initiated the topic of multi-parameter stochastic processes, introducing in particular the multi-parameter Brownian motion.

As should be quite clear, the account given here does not describe the whole of Paul Lévy's mathematical achievements, and one can consult for many more details the first paper (by Michel Loève) published in the first issue of the *Annals of Probability* (1973). It also does not account for the humanity and gentleness of the person Paul Lévy. But I would like to end this short exposition of Paul Lévy's work by hoping that this series will contribute to fulfilling the program, which he initiated.

Paris, France   Jean Jacod

Fractional Lévy Fields

Serge Cohen

Abstract In this survey, we would like to summarize most of the results concerning the so-called fractional Lévy fields in a way as self-contained as possible. Beside the construction of these fields, we are interested in the regularity of their sample paths, and self-similarity properties of their distributions. It turns out that for applications, we often need non-homogeneous fields that are only locally self-similar. Then we explain how to identify those models from a discrete sample of one realization of the field. At last some simulation techniques are discussed.

Mathematics Subject Classification 2000: Primary: 60G10 , 60G70 , 60J10
Secondary: 91B28, 91B84

Keywords Fractional fields • Lévy random measure • Random fields

1 Introduction

Irregular phenomena appear in various fields of scientific research: fluid mechanics, image processing and financial mathematics, for example. Experts in those fields often ask mathematicians to develop models, both easy to use, and relevant for their applications. In this perspective, fractional fields are very often used to model irregular phenomena. Among the huge literature devoted to the topic, one can refer the reader to [1, 11, 12] for an overview of fractional fields for applications and of related works.

One of the simplest model is the fractional Brownian motion introduced in [19], and further developed in [25]. Simulation of the fractional Brownian motion is

S. Cohen (✉)
Institut de Mathématiques de Toulouse, Université Paul Sabatier, Université de Toulouse, 118 Route de Narbonne 31062 Toulouse Cedex 9, France
e-mail: Serge.Cohen@math.univ-toulouse.fr; http://www.math.univ-toulouse.fr/~cohen/

S. Cohen et al., *Lévy Matters II*, Lecture Notes in Mathematics 2061,
DOI 10.1007/978-3-642-31407-0_1, © Springer-Verlag Berlin Heidelberg 2012

now both theoretically and practically well understood (see [5] for a survey on this problem). Many other fractional fields with heavy tailed marginals have been proposed for applications, see Chap. 7 in [34] for an introduction to fractional stable processes. More recently other processes that are neither Gaussian nor stable have been proposed to model Internet traffic (cf. [10, 17]). The common feature for many of these fields, see also [3, 7, 21], is the fact that they are obtained by a stochastic integration of a deterministic kernel with respect to some random measure. In terms of models, we can think that the probabilistic structure of the irregular phenomena (light or heavy tails for instance) is implemented in the random measure and the correlation structure is built in the deterministic kernel. Engineers will have to try many kernels and random measures before finding the more appropriate one for their applications.

We start this survey with a reminder of the integration of deterministic function with respect to random measures. Our goal in Sect. 2 is to provide a self-contained introduction to Lévy random measures on $\mathbb{R}^d$, which are used to construct fractional fields. If you already know this topic you can skip this part. If it is not the case, we recall also Poisson random measures that are a basic tool for the construction of Lévy random measures, and stable random measures, which are a limit case. In Sect. 3, we consider stable fractional fields, starting from Gaussian fractional fields. Actually the Gaussian case (including the celebrated fractional Brownian motion) is an historical motivation to introduce other fractional fields. We also consider in this part fractional stable fields, which are non-Gaussian, and which are also self-similar. To have additional information on these fields, we refer the interested reader to [34]. In Sect. 4 we study the main topic of this survey: Fractional Lévy fields which are neither Gaussian nor stable. They enjoy interesting properties for modelization purposes that make them an intermediate case between stable and Gaussian fractional fields. On the one hand, they inherit from their Gaussian ancestors second order moments and they have actually the same covariance structure as fractional Brownian fields. On the other hand, their sample paths have very different regularities, when they are of moving average type and of harmonizable type. Moreover they are intermediate between Gaussian and Stable fields in the sense of the asymptotic self-similarity, which is also defined in this section. Roughly speaking scaling limits at infinity or at zero of Lévy fractional fields are sometimes Gaussian or stable fractional fields. All these properties are studied in details in this section, which is concluded by a generalization of fractional fields to multifractional fields. Basically "fractional" means that a single index $0 < H < 1$ describes the distribution of the sample paths. Sometime for application, one would like to have the index H dependent on the location, and in this case, the index becomes a function. It is the main motivation for "multifractional" Lévy fields. An important feature of (multi)fractional Lévy fields is the fact that the fractional index or the multifractional function can be estimated asymptotically with only one discretized observation of a sample path. It is a very convenient property of these models, which is explained in Sect. 5, and, which helps engineers to calibrate the parameters to mimic their experimental data. In Sect. 6 a generic simulation technique is provided to have computer simulations of the fields introduced in the previous section, and it helps to understand qualitatively fractional fields.

2 Random Measures

2.1 Poisson Random Measure

In this article, we are using Poisson measures to build stable and Lévy measures. A Poisson random measure N on a measurable state space $(S, \mathscr{S})$, endowed with deterministic measure n, is an independently scattered σ-additive set function defined on $\mathscr{S}_0 = \{A \in \mathscr{S} \text{ s. t. } n(A) < +\infty\}$. Thus, for $A \in \mathscr{S}_0$, the random variable $N(A)$ has a Poisson distribution with mean $n(A)$

$$\mathbf{P}(N(A) = k) = e^{-n(A)} \frac{(n(A))^k}{k!}$$

for $k = 0, 1, 2, \ldots$ The measure n is called the mean measure of N. Moreover the Poisson random measure N is independently scattered in the following sense.

Definition 2.1. A random measure M is independently scattered, if for a finite number of sets $\{A_i\}_{j \in J}$ that are pairwise disjoint, the random variables $(M(A_j))_{j \in J}$ are independent.

Furthermore the Poisson random variables have expectation and variance given by their mean i.e.

$$\mathbb{E}(N(A)) = \mathrm{var} N(A) = n(A)$$

for every $A \in \mathscr{S}_0$. Their characteristic function is easily computed

$$\mathbb{E}(e^{iv N(A)}) = \exp\left(n(A)(e^{iv} - 1)\right)$$

for every $v \in \mathbb{R}$.

Our goal is to build Lévy (symmetric or complex isotropic) random measures $M(ds)$ to define fractional Lévy fields. The first step in that direction is to define a stochastic integral with respect to Poisson measures. In this article, the Poisson measures used to build Lévy random measure are compensated. Let us define compensated Poisson measures on $\mathbb{R}^d \times \mathbb{R}$ with mean measure $n(ds, du)$

$$\widetilde{N} = N - n.$$

The additional variable u is used in Sect. 2.2 and its meaning will be explained there.

Then for every function $\varphi : \mathbb{R}^d \times \mathbb{R} \to \mathbb{R}$ such that $\varphi \in L^2(\mathbb{R}^d \times \mathbb{R}, n)$ the stochastic integral

$$\int_{\mathbb{R}^d \times \mathbb{R}} \varphi(s, u) \widetilde{N}(ds, du) \qquad (1)$$

is defined as the limit in $L^2(\Omega)$ of

$$\int_{\mathbb{R}^d \times \mathbb{R}} \varphi_k(s, u) \widetilde{N}(ds, du),$$

where φ_k is a simple function of the form $\sum_{j \in J} a_j \mathbf{1}_{A_j}$, where J is finite. Then

$$\int_{\mathbb{R}^d \times \mathbb{R}} \sum_{j \in J} a_j \mathbf{1}_{A_j} \widetilde{N}(ds, du) \overset{def}{=} \sum_{j \in J} a_j \widetilde{N}(A_j),$$

where the Poisson random variables $N(A_j)$ have intensity $n(A_j)$ and are independent since the sets A_j are supposed pairwise disjoint. Consequently the characteristic function $\Phi(v)$ of $\sum_{j \in J} a_j \widetilde{N}(A_j)$ is

$$\Phi(v) = \exp\left[\sum_{j \in J} n(A_j)(e^{ia_j v} - 1 - ia_j v)\right],$$

and the convergence in $L^2(\Omega)$ implies that the characteristic function of the stochastic integral is $\forall v \in \mathbb{R}$

$$\mathbb{E} \exp\left(iv \int \varphi d\widetilde{N}\right) = \exp\left[\int_{\mathbb{R}^d \times \mathbb{R}} [\exp(iv\varphi) - 1 - iv\varphi] n(ds, du)\right]. \qquad (2)$$

Moreover $\mathrm{var} N(A) = n(A)$ yields

$$\mathbb{E}\left(\int_{\mathbb{R}^d \times \mathbb{R}} \varphi(s, u) \widetilde{N}(ds, du)\right)^2 = \int_{\mathbb{R}^d \times \mathbb{R}} \varphi^2(s, u) n(ds, du), \qquad (3)$$

first for simple functions, then for every $\varphi(s, u) \in L^2(n(ds, du))$.

Please note that as Poisson random measure, compensated Poisson random measure are independently scattered.

We will extend the definition of the integral of a function φ with respect to a compensated Poisson measure in Sect. 6.2 when $\varphi \notin L^2$.

2.2 Lévy Random Measure

With the help of the compensated Poisson random measure, we will define Lévy random measure for which the control measure has a finite moment of order 2.

It is not the case of stable Lévy measure, but the random stable measures are defined as limits of such Lévy measures, which are also useful to construct locally self-similar fields considered in the Sect. 4. Let us take a control measure μ such that

$$\int_{\mathbb{R}} |u|^p \mu(du) < \infty \quad \forall p \geq 2. \tag{4}$$

Then, a Poisson random measure $\widetilde{N}$ is associated with the mean measure $n(ds, du) = ds\mu(du)$, and for every function $f \in L^2(\mathbb{R}^d, ds)$ one can define

$$\int_{\mathbb{R}^d} f(s)M(ds) \overset{def}{=} \int_{\mathbb{R}^d \times \mathbb{R}} f(s)u\widetilde{N}(ds, du), \tag{5}$$

since $f(s)u \in L^2(\mathbb{R}^d, ds\mu(du))$.

Actually the name "Lévy" comes from the fact that in dimension $d = 1$, this random Lévy measure is related to the increments of processes with stationary and independent increments, which are called Lévy processes.

Let us recall the Lévy Kintchine formula (See for instance [29] Theorem 42 and 43), which yields the characteristic function of a Lévy process $(X_t)_{t\in\mathbb{R}}$ with Lévy measure μ:

$$\forall t \in \mathbb{R}, \quad \mathbb{E}(\exp(ivX_t)) = \exp(-t\psi(v)), \tag{6}$$

where $\forall v \in \mathbb{R}$

$$\psi(v) = \frac{\sigma^2}{2}v^2 - ibv + \int_{|x|\geq 1}(1 - e^{ivx})\mu(dx) + \int_{|x|<1}(1 - e^{ivx} + ivx)\mu(dx). \tag{7}$$

In (7), the terms on the right hand correspond to a decomposition of the Lévy process itself in three processes: a Brownian motion with variance σ^2, a deterministic process, and a "jump" process, which corresponds to the last two integrals. In particular the Lévy measure μ in (7) satisfies $\int \inf(1, x^2)\mu(dx) < \infty$, and can be viewed as the expectation of random measure related to the jumps of X. Last every infinitely divisible random variable can be obtained as the marginal at time $t = 1$ of a Lévy process.

Then one can easily compute the characteristic function of the random variable defined by the integral in (5) and check that this variable is infinitely divisible.

Moreover one can consider heuristically $u\widetilde{N}(ds, du)$ as the derivative of some Lévy fields. If you agree with this picture, and you assume that the random measure $\widetilde{N}(ds, du)$ contains a Dirac mass at point $(s(\omega), u(\omega))$, then you may think of $u(\omega)$ as the "jump" of the Lévy field at point $s(\omega)$.

To illustrate further the relationship between Lévy measure and Lévy processes, let us remark that the process

$$X(t) = \int_{\mathbb{R}} \mathbf{1}_{[0,t]}(s)M(ds), \ t \geq 0$$

is a Lévy process. Since the Poisson measures are independently scattered, the process has independent increments. The mean measure of the compensated Poisson measure used to construct Lévy or stable random measures is the Lebesgue measure, the characteristic function (2) implies the stationarity of the increments. Moreover, one can deduce from the Lévy Kintchine formula that the (non-random) Lévy measure of the Lévy process is μ.

Let us now state an isometry property for Lévy random measure:

$$\mathbb{E}\left(\int_{\mathbb{R}^d} f(s)M(ds)\right)^2 = \int_{\mathbb{R}} u^2\mu(du)\|f\|_{L^2(\mathbb{R}^d)}^2, \tag{8}$$

which is a consequence of (3). When A is a Borel set of $\mathbb{R}^d$ with finite Lebesgue measure, we will denote by $M(A)$ the random variable $M(\mathbf{1}_A)$.

Please note that Lévy random measure are independently scattered in the sense of Definition 2.1.

2.3   Real Stable Random Measure

In this article we will also consider the special instance of stable random variables (or stable measures). Let us recall characteristic functions of symmetric α-stable random variables X

$$\mathbb{E}(e^{ivX}) = \exp\left(-\sigma^\alpha|v|^\alpha\right),$$

where $0 < \alpha < 2$, and where σ is called the scale factor. Please note that the case $\alpha = 2$ formally corresponds to the Gaussian case. Because of harmonizable representations, we will also need complex valued stable variables. Let us introduce complex isotropic stable variables $X_1 + iX_2$, their characteristic functions are given by:

$$\mathbb{E}(e^{i(v_1X_1+v_2X_2)}) = \exp\left(-\sigma_\alpha(v_1^2 + v_2^2)^{\alpha/2}\right).$$

At this point, the α-stable symmetric random measure M_α will be defined as a limit of Lévy random measures $M_{\alpha,R}(ds)$. Let us consider $f \in L^\alpha(\mathbb{R}^d) \cap L^2(\mathbb{R}^d)$, then, for any positive number R, the integral on the left hand side

$$\int_{\mathbb{R}^d} f(s)M_{\alpha,R}(ds) \stackrel{def}{=} \int_{\mathbb{R}^d \times \mathbb{R}} f(s)u\widetilde{N_R}(ds, du) \tag{9}$$

is defined by the integral on the right hand side, where the mean measure of N_R is $n_R(ds, du) = \frac{1_{|u| \le R}}{|u|^{1+\alpha}} ds du$. Roughly speaking, $M_{\alpha,R}$ is a stable measure where the "big jumps" have been truncated. Let us now check that $\int_{\mathbb{R}^d} f(s)M_{\alpha,R}(ds)$ converges in distribution, when $R \to +\infty$, to a symmetric α-stable random variable, which scale factor is given by $\left(\int_{\mathbb{R}^d} |f(s)|^\alpha ds\right)^{1/\alpha}$. The limit will be denoted by

$\int_{\mathbb{R}^d} f(s)M_\alpha(ds)$. The random measure $M_\alpha(ds)$ is called a random stable measure and we shall only consider in this article symmetric random stable measure. Let us give a formal definition of a symmetric random stable measure (in short SαS random measure).

Definition 2.2. Let $(E, \mathscr{E}, m)$ be a measurable space with a sigma finite measure m. A symmetric random stable measure M_α satisfies that for every $\mathscr{E}$-measurable function f such that $\int_E |f|^\alpha dm < \infty$, $\int_E f(s)M_\alpha(ds)$ is a symmetric random variable with scale factor $\left(\int_E |f|^\alpha dm\right)^{1/\alpha}$. Moreover M_α is assumed to be independently scattered.

In this article, symmetric random stable measure will be limit of Lévy random measure.

Lemma 2.3. *For all functions $f \in L^\alpha(\mathbb{R}^d) \cap L^2(\mathbb{R}^d)$, the limit in distribution*

$$\lim_{R \to +\infty} \int_{\mathbb{R}^d} f(s)M_{\alpha,R}(ds)$$

exists. Moreover, the limit in distribution is the distribution of an α-stable random variable with scale factor $\left(C(\alpha)\int_{\mathbb{R}^d} |f(s)|^\alpha ds\right)^{1/\alpha}$, where

$$C(\alpha) = \frac{\pi}{2\alpha\Gamma(\alpha)\sin(\frac{\pi\alpha}{2})}, \tag{10}$$

where Γ is the classical Gamma function.

Proof. Because of (2),

$$-\log\left(\mathbb{E}\exp\left(iv\int f(s)M_{\alpha,R}(ds)\right)\right)$$

$$= \int_{\mathbb{R}^d \times \mathbb{R}} [1 - \exp(ivf(s)u) + ivf(s)u]\,ds\,\frac{\mathbf{1}_{|u|\le R}du}{|u|^{1+\alpha}}. \tag{11}$$

Moreover, for $0 < \alpha < 2$,

$$C(\alpha)|x|^\alpha = \int_{\mathbb{R}} [1 - e^{ixu} + ixu\mathbf{1}_{|u|\le R}]\frac{du}{|u|^{1+\alpha}}, \tag{12}$$

for every $x \in \mathbb{R}$, where the constant $C(\alpha)$ is given by letting $x = 1$ in (12) and it does not depend on R. Hence

$$C(\alpha) = \int_{\mathbb{R}} \frac{(1 - \cos(u))}{|u|^{1+\alpha}}du$$

$$= \frac{\pi}{2\alpha\Gamma(\alpha)\sin(\frac{\pi\alpha}{2})}.$$

This formula for $C(\alpha)$ can be deduced for instance from 7.2.13 p. 328 in [34]. Hence,

$$C(\alpha)|v|^\alpha \int_{\mathbb{R}^d} |f(s)|^\alpha \, ds = \int_{\mathbb{R}^d} ds \int_{\mathbb{R}} [1 - e^{if(s)vu} + if(s)vu\mathbf{1}_{|u|\leq R}] \frac{du}{|u|^{1+\alpha}}. \quad (13)$$

Then,

$$-C(\alpha)|v|^\alpha \int_{\mathbb{R}^d} |f(s)|^\alpha \, ds - \log\left(\mathbb{E}\exp\left(iv\int f(s)M_{\alpha,R}(ds)\right)\right)$$
$$= -\int_{\mathbb{R}^d} ds \int_{\mathbb{R}} \frac{du}{|u|^{1+\alpha}} (1 - e^{if(s)vu})\mathbf{1}_{|u|>R}.$$

Please note that the last line is negative and finite for any $R > 0$. Hence, it converges to 0 by monotone convergence. Consequently the convergence in distribution is established and

$$\log\left(\mathbb{E}\exp\left(iv\int f(s)M_\alpha(ds)\right)\right) = -C(\alpha)|v|^\alpha \int_{\mathbb{R}^d} |f(s)|^\alpha \, ds. \quad (14)$$

$\square$

Definition 2.4. For all functions $f \in L^\alpha(\mathbb{R}^d) \cap L^2(\mathbb{R}^d)$, we denote by $\int_{\mathbb{R}^d} f(s) M_\alpha(ds)$

$$\lim_{R\to+\infty} \int_{\mathbb{R}^d} f(s)M_{\alpha,R}(ds).$$

Actually $\int_{\mathbb{R}^d} f(s)M_\alpha(ds)$ is defined for every function $f \in L^\alpha(\mathbb{R}^d)$ since $L^\alpha(\mathbb{R}^d) \cap L^2(\mathbb{R}^d)$ is dense in $L^\alpha(\mathbb{R}^d)$. At this point it is not clear that $\int_{\mathbb{R}^d} f(s)M_\alpha(ds)$ does define an α-symmetric stable random measure in the sense of Definition 2.2. But the integration with respect to $M_\alpha(ds)$ could be defined as a stochastic field $(\int_{\mathbb{R}^d} f(s)M_\alpha(ds))_{f\in L^\alpha}$ parameterized by L^α as in Sect. 3.2 in [34], since the proof of Lemma 2.3 can be carried for any finite number of functions $f_1, \ldots, f_k \in L^\alpha(\mathbb{R}^d)$. Moreover the independent scattering property is a consequence of the same property for Lévy random measures, which is a direct consequence of the independent scattering for random Poisson measures. We omit the complete proof but we claim that there exists an α-symmetric stable random measure $M_\alpha(ds)$ in the sense of Definition 2.2 such that $\lim_{R\to+\infty} \int_{\mathbb{R}^d} f(s)M_{\alpha,R}(ds) \overset{(d)}{=} \int_{\mathbb{R}^d} f(s)M_\alpha(ds)$.

2.4 Complex Isotropic Random Measure

In the Sect. 3, we have to construct a complex isotropic α-stable random measure to define a self-similar field. Let us sketch the various steps that allow us to define this random measure starting from a complex Poisson random measure. Since this construction is parallel to the one in the real case, we do not give details.

Let us consider an isotropic control measure on the complex field $\mu(dz)$. Isotropy means that, if P is the map $P(\rho \exp(i\theta)) = (\theta, \rho) \in [0, 2\pi) \times (0, \infty)$, then

$$P(\mu(dz)) = d\theta \mu_\rho(d\rho) \tag{15}$$

where $d\theta$ is the uniform measure on $[0, 2\pi)$. We still assume that

$$\int_{\mathbb{C}} |u|^p \mu(du) < \infty \quad \forall p \geq 2, \tag{16}$$

and we consider the associated Poisson random measure $\tilde{N} = N - n$, where N is a Poisson random measure on $\mathbb{R}^d \times \mathbb{C}$ with mean measure $n(d\xi, dz) = d\xi \mu(dz)$. Similarly to the real case, one can define a stochastic integral for every complex valued function $\varphi \in L^2(\mathbb{C} \times \mathbb{R}^d, d\xi \mu(dz))$ denoted by

$$\int_{\mathbb{R}^d \times \mathbb{C}} \varphi(\xi, z) \tilde{N}(d\xi, dz). \tag{17}$$

If φ is real valued so is $\int \varphi d\tilde{N}$, and if $\Re(z)$ denotes the real part of a complex z, then $\Re(\int \varphi d\tilde{N}) = \int \Re(\varphi) d\tilde{N}$, the same property is true for the imaginary part $\Im$ of stochastic integrals. Furthermore, for every u, $v \in \mathbb{R}$,

$$\mathbb{E} \exp \left(i(u \int \Re(\varphi) d\tilde{N} + v \int \Im(\varphi) d\tilde{N}) \right)$$

$$= \exp[\int_{\mathbb{R}^d \times \mathbb{C}} [\exp(i(u\Re(\varphi) + v\Im(\varphi))) - 1 - i(u\Re(\varphi) + v\Im(\varphi))] d\xi \mu(dz)]. \tag{18}$$

Like real random Poisson measures, complex Poisson measures satisfy an isometry property

$$\mathbb{E} \left| \int_{\mathbb{R}^d \times \mathbb{C}} \varphi(\xi, z) \tilde{N}(d\xi, dz) \right|^2 = \int_{\mathbb{R}^d \times \mathbb{C}} |\varphi(\xi, z)|^2 n(d\xi, dz), \tag{19}$$

where $|\varphi(\xi, z)|$ is the complex modulus of the complex number $\varphi(\xi, z)$. This property is obvious for simple functions of the form $\sum_{j \in J} a_j \mathbf{1}_{A_j}$ and is trivially extended to functions in L^2 by letting a sequence of simple functions converges to every L^2 function. One can now define the complex isotropic Lévy random measure:

Definition 2.5.

$$\int_{\mathbb{R}^d} f(\xi) M(d\xi) = \int_{\mathbb{R}^d \times \mathbb{C}} [f(\xi)z + f(-\xi)\bar{z}] \tilde{N}(d\xi, dz) \tag{20}$$

for every function $f : \mathbb{R}^d \to \mathbb{C}$ where $f \in L^2(\mathbb{R}^d)$.

One can first state some elementary properties of the integral defined in (20). When

$$\forall \xi \in \mathbb{R}^d, \quad f(-\xi) = \overline{f(\xi)}, \tag{21}$$

the following integral is a real number since

$$\int_{\mathbb{R}^d} f(\xi) M(d\xi) = \int_{\mathbb{R}^d \times \mathbb{C}} 2\Re(f(\xi)z) \tilde{N}(d\xi, dz) \tag{22}$$

$$= 2\Re \left(\int_{\mathbb{R}^d \times \mathbb{C}} f(\xi) z \tilde{N}(d\xi, dz) \right). \tag{23}$$

When f satisfies (21), for every measurable odd function a,

$$\int f(\xi) \exp(ia(\xi)) M(d\xi) \overset{(d)}{=} \int f(\xi) M(d\xi) \tag{24}$$

is a consequence of (18). Moreover stochastic integrals have symmetric distributions:

$$-\int f(\xi) M(d\xi) \overset{(d)}{=} \int f(\xi) M(d\xi). \tag{25}$$

Then an isometry property for the Lévy measure $M(d\xi)$, when f satisfies (21),

$$\mathbb{E} |\int_{\mathbb{R}^d} f(\xi) M(d\xi)|^2 = 4\pi \|f\|^2_{L^2(\mathbb{R}^d)} \int_0^{+\infty} \rho^2 \mu_\rho(d\rho) \tag{26}$$

holds, it is a consequence of (19). The properties of complex valued random Lévy measures are proved below

1. Let us compute the characteristic function of $\int f(\xi) M(d\xi)$, which is denoted by

$$\Phi(u, v) = \mathbb{E} \exp \left(i(u\Re \int f(\xi) M(d\xi) + v\Im \int f(\xi) M(d\xi)) \right).$$

Because of Definition 2.5, and of (18),

$$\Phi(u, v) = \mathbb{E} \exp \left(i(u \int \Re(f(\xi)z + f(-\xi)\bar{z}) \tilde{N}(d\xi, dz) \right.$$

$$+ v \int \Im(f(\xi)z + f(-\xi)\bar{z}) \tilde{N}(d\xi, dz)) \right)$$

$$= \exp \left(\int_{\mathbb{R}^d \times \mathbb{C}} [\exp(g_{u,v}(\xi, z)) - 1 - g_{u,v}(\xi, z)] d\xi d\mu(z) \right),$$

where $g_{u,v}(\xi, z)) = i(u\Re(f(\xi)z + f(-\xi)\bar{z}) + v\Im(f(\xi)z + f(-\xi)\bar{z})$. Then,

$$\log(\Phi(u, v)) = \int_{\mathbb{R}^d \times [0,2\pi] \times \mathbb{R}_*^+} [\exp(g_{u,v}(\xi, \rho e^{i\theta})) - 1 - g_{u,v}(\xi, \rho e^{i\theta})] d\xi d\theta \mu_\rho(d\rho).$$

If we replace f by $-f$ in $g_{u,v}$, one can rewrite the product $-f(\xi)\rho e^{i\theta} = f(\xi)\rho e^{i(\theta+\pi)}$. Hence, the characteristic function of $\int f(\xi)M(d\xi)$ is the same as the characteristic function of $-\int f(\xi)M(d\xi)$, because of the invariance of the measure $d\theta$ with respect to the translation of magnitude π.

2. Please note that both integrals $-\int f(\xi)M(d\xi)$ and $\int f(\xi)\exp(ia(\xi))M(d\xi)$ are real valued. Then the log-characteristic function of $\int f(\xi)\exp(ia(\xi))M(d\xi)$

$$\Phi(u) = \log\left(\mathbb{E}\exp\left(i(u\int f(\xi)\exp(ia(\xi))M(d\xi))\right)\right)$$

is given by

$$\int_{\mathbb{R}^d \times [0,2\pi] \times \mathbb{R}_*^+} [\exp(iu2\Re(f(\xi)e^{ia(\xi)}\rho e^{i\theta})) - 1$$
$$- iu2\Re(f(\xi)e^{ia(\xi)}\rho e^{i\theta})]d\xi d\theta \mu_\rho(d\rho).$$

It is equal to $\log\left(\mathbb{E}\exp\left(i(u\int f(\xi)M(d\xi))\right)\right)$ because of the invariance of the measure $d\theta$ with respects to the translation of magnitude $a(\xi)$.

3. Because of (19)

$$\mathbb{E}|\int_{\mathbb{R}^d} f(\xi)M(d\xi)|^2 = \mathbb{E}|\int_{\mathbb{R}^d \times \mathbb{C}} 2\Re(f(\xi)z)\tilde{N}(d\xi, dz)|^2$$
$$= 4\int_{\mathbb{R}^d \times \mathbb{C}} \Re(f(\xi)\rho e^{i\theta})^2 d\xi \mu(d\rho)d\theta$$
$$= 4\int_0^{+\infty} \rho^2 \mu_\rho(d\rho) \int_{\mathbb{R}^d} |f(\xi)|^2 d\xi \int_{[0,2\pi]} \Re(e^{i\theta})^2 d\theta$$
$$= 4\pi\|f\|_{L^2(\mathbb{R}^d)}^2 \int_0^{+\infty} \rho^2 \mu_\rho(d\rho).$$

Hence we get (26).

Actually the characteristic function (18) allows us to compute every moments of the stochastic integrals $\int f(\xi)M(d\xi)$.

Proposition 2.6. *If $f \in L^2(\mathbb{R}^d) \bigcap L^{2p}(\mathbb{R}^d)$, and f satisfies (21), then $\int f(\xi)M(d\xi)$ is in $L^{2p}(\Omega)$. Moreover*

$$\mathbb{E}\left((\int f(\xi)M(d\xi))^{2p}\right) = \sum_{n=1}^{p}(2\pi)^n$$

$$\sum_{P_n}\prod_{q=1}^{n}\frac{(2m_q)!\|f\|_{2m_q}^{2m_q}\int_0^{+\infty}\rho^{2m_q}\mu_\rho(d\rho)}{(m_q!)}, \qquad (27)$$

where $\sum_{P_n}$ stands for the sum over the set of partitions P_n of $\{1,\ldots,2p\}$ in n subsets K_q such that the cardinality of K_q is $2m_q$, with $m_q \geq 1$, and, where $\|f\|_{2m_q}$ is the $L^{2m_q}(\mathbb{R}^d)$ norm of f.

Proof. An expansion in power series of both sides of (18) yields the result. Let us sketch the beginning of the computations. We can write

$$\mathbb{E}\exp\left(i(u\int f(\xi)M(d\xi))\right) = \exp\left(\int_{\mathbb{R}^d\times\mathbb{C}}[\exp(iu2\Re(f(\xi)z)) - 1\right.$$

$$\left. - iu2\Re(f(\xi)z)]d\xi\mu(dz)\right)$$

$$= \exp\left(\int_{\mathbb{R}^d\times\mathbb{C}}\sum_{k=2}^{\infty}\frac{(iu2\Re(f(\xi)z))^k}{k!}d\xi\mu(dz)\right)$$

$$= \sum_{l=0}^{\infty}\left(\sum_{k=2}^{\infty}\int_{\mathbb{R}^d\times\mathbb{C}}\frac{(iu2\Re(f(\xi)z))^k d\xi\mu(dz)}{k!}\right)^l/l!$$

and we get on the right hand side (after tedious computations) the coefficient of u^n, which is the nth-moment of $\int f(\xi)M(d\xi)$. It yields eventually (27).   □

One can also define a complex isotropic α-stable random measure as a limit of complex isotropic Lévy random measures. Let us consider a complex isotropic random Lévy measures $M_{\alpha,R}$ associated to $n_R(d\xi, dz) = d\xi\frac{dz}{|z|^{1+\alpha}}\mathbf{1}_{|z|\leq R}$, one can prove a lemma similar to Lemma 2.3.

Lemma 2.7. *For all functions $f \in L^\alpha(\mathbb{R}^d)\cap L^2(\mathbb{R}^d)$, that satisfy (21), the limit in distribution of:*

$$\lim_{R\to+\infty}\int_{\mathbb{R}^d}f(\xi)M_{\alpha,R}(d\xi) \qquad (28)$$

exists. Moreover the limit in distribution is a real valued α-stable random variable with scale factor

$$\left(C(\alpha)2^{\alpha-1}\int_0^{2\pi}|\cos(\theta)|^\alpha d\theta\int_{\mathbb{R}^d}|f(\xi)|^\alpha d\xi\right)^{1/\alpha},$$

where $C(\alpha)$ is defined in (10).

Proof. Let consider for $u \in \mathbb{R}$ the characteristic function

$$- \log \left(\mathbb{E} \exp \left(iu \int f(\xi) M_{\alpha,R}(d\xi) \right) \right)$$

$$= \int_{\mathbb{R}^d \times [0,2\pi] \times (0,+\infty)} [\exp(iu\rho 2\Re(f(\xi)e^{i\theta})) - 1 - iu\rho 2\Re(f(\xi)e^{i\theta})] \mathbf{1}_{|\rho| \leq R} d\xi d\theta \frac{d\rho}{\rho^{1+\alpha}}.$$

As in Lemma 2.3, one can show that

$$\lim_{R \to \infty} - \log \left(\mathbb{E} \exp \left(iu \int f(\xi) M_{\alpha,R}(d\xi) \right) \right)$$

$$= \frac{C(\alpha)}{2} |u|^\alpha \int_{\mathbb{R}^d} \int_0^{2\pi} |2\Re(f(\xi)e^{i\theta})|^\alpha d\xi d\theta,$$

which can also be written

$$C(\alpha) 2^{\alpha-1} \int_0^{2\pi} |\cos(\theta)|^\alpha d\theta \int_{\mathbb{R}^d} |f(\xi)|^\alpha d\xi.$$

$\square$

Definition 2.8. For all functions $f \in L^\alpha(\mathbb{R}^d) \cap L^2(\mathbb{R}^d)$, we denote by $\int_{\mathbb{R}^d} f(\xi) M_\alpha(d\xi)$

$$\lim_{R \to +\infty} \int_{\mathbb{R}^d} f(\xi) M_{\alpha,R}(d\xi).$$

Please note that this definition can be extended to $f \in L^\alpha(\mathbb{R}^d)$ in a similar way as in Sect. 2.3.

3 Stable Fields

3.1 Gaussian Fields

Historically, the most celebrated fractional process is the fractional Brownian motion introduced in [19, 25]. It is a Gaussian process and a very large literature is devoted to Gaussian fractional fields. See [9] for an introduction to Gaussian fractional fields. In this survey the focus is on non-Gaussian fractional fields, but since most of the concepts are motivated by Gaussian fields, we have to recall some facts concerning fractional Brownian motion starting with the definition.

Definition 3.1. Fractional Brownian fields (in short fBf) are centered Gaussian fields and their covariances are given by

$$R(x, y) = \frac{C}{2}\{\|x\|^{2H} + \|y\|^{2H} - \|x - y\|^{2H}\}, \tag{29}$$

where $C > 0$.

Hence the word "fractional" can be related to the covariance structure of the fields in the sense that a fractional power is used to define the covariance. But this way of thinking is a bit shallow, and not easily extended to fields, which are not square integrable. To go further, one can use integrable representations of fractional Brownian fields, which are related to some "fractional" integration, which is a deeper explanation of the vocabulary. Actually there exist two representations. First the moving average representation of fractional Brownian fields

$$B_H(t) \overset{(d)}{=} \int_{\mathbb{R}^d} \left[\|t - s\|^{H-d/2} - \|s\|^{H-d/2} \right] W(ds), \tag{30}$$

for $0 < H < 1$; $H \neq \frac{d}{2}$. The notation $(X(t))_{t\in\mathbb{R}^d} \overset{(d)}{=} (Y(t))_{t\in\mathbb{R}^d}$ means that for every $n \geq 1$ and $t_1, \ldots, t_n \in \mathbb{R}^d$

$$(X_{t_1}, \ldots, X_{t_n}) \overset{(d)}{=} (Y_{t_1}, \ldots, Y_{t_n}). \tag{31}$$

We also have the harmonizable representation

$$X(x) \overset{(d)}{=} \int_{\mathbb{R}^d} \frac{e^{-ix.\xi} - 1}{\|\xi\|^{H+d/2}} \widehat{W}(d\xi), \tag{32}$$

where both representations are using random Brownian measures, and we refer the reader to the Chap. 2 in [9] for definitions and discussions of these random measures. Since we will use other measures in this survey, we will not give details here on this topic. Let us also recall some properties of the fractional Brownian fields that they share with some non-Gaussian fractional fields. Fractional Brownian fields have homogeneous distributions at least in two senses. First they are with stationary increments, which makes them easier to study from a theoretical point of view, but which can be a drawback, when one wants to model data that are not stationary. Anyway, let us a give a definition of fields with stationary increments.

Definition 3.2. A field $(X(t))_{t\in\mathbb{R}^d}$ such that $\forall s \in \mathbb{R}^d$

$$(X(t + \delta) - X(s + \delta))_{t\in\mathbb{R}^d} \overset{(d)}{=} (X(t) - X(s))_{t\in\mathbb{R}^d} \tag{33}$$

is called with stationary increments.

Second, fractional Brownian fields are self-similar with index H, which means that

$$(X(\varepsilon x))_{x\in\mathbb{R}^d} \overset{(d)}{=} \varepsilon^H(X(x))_{x\in\mathbb{R}^d}. \tag{34}$$

Here again we have a fractional power and another motivation for the word fractional. Moreover sample paths of fractional Brownian fields enjoy also almost surely regularity properties.

Theorem 3.3. *For every $H' < H$ there exists a modification of a fBf B_H such that*

$$\mathbb{P}\left(\sup_{\substack{|s-t|<\epsilon(\omega)\\|s|\leq 1,\,|t|\leq 1}} \left(\frac{B_H(s) - B_H(t)}{|s-t|^{H'}}\right) \leq \delta\right) = 1, \tag{35}$$

where ϵ is positive random variable and $\delta > 0$. Moreover the pointwise Hölder exponent for every $t \in \mathbb{R}$

$$\sup\{H', \lim_{\epsilon\to 0} \frac{B_H(t+\epsilon) - B_H(t)}{|\epsilon|^{H'}} = 0\} = H \tag{36}$$

almost surely.

First we recall the classical definition of a modification and we don't go in further details on this point.

Definition 3.4. Modification.
A field Y is a modification of a field X if, for every t, $\mathbb{P}(X_t = Y_t) = 1$.

The first part of the theorem claims that the sample paths of the fractional Brownian motion are almost surely locally Hölder continuous for every $H' < H$, and (36) means that the sample paths are not $H-$ Hölder continuous. Therefore some definitions are introduced.

Definition 3.5. Let $f : \mathbb{R}^d \mapsto \mathbb{R}$ be a function such that there exist $C > 0$ and $0 < H < 1$ so that

$$|f(t) - f(u)| \leq C\|t - u\|^H \quad \forall t,\, u \in \mathbb{R}^d.$$

The function f is called H-Hölder continuous. The set of H-Hölder continuous functions on $[0, 1]^d$ is denoted by $\mathscr{C}^H$.

Unfortunately the conclusion of many theorems like Theorem 3.3 is only that the sample paths are locally Hölder continuous, which means that their restrictions to every compact sets is H-Hölder continuous.

Definition 3.6. Let $f : \mathbb{R}^d \mapsto \mathbb{R}$ be a function such that on every compact set K there exist a constant $C(K) > 0$ depending only on K, and $0 < H < 1$ so that

$$|f(t) - f(u)| \leq C(K)\|t - u\|^H \quad \forall t, \ u \in K.$$

The function f is called locally H-Hölder continuous.

Please note that if $H' < H$, H-Hölder continuous functions are H'-Hölder continuous. The same property is true locally. Hence if we know that a function is H-Hölder continuous, one can wonder what is the best H. These considerations lead to the definition of Hölder exponents that can have different natures: global, local or pointwise. In this survey we will only be concerned with pointwise Hölder exponents.

Definition 3.7. A real valued function f defined in a neighborhood of t, has a pointwise Hölder exponent H if

$$H(x) = \sup\{H', \quad \lim_{\|\epsilon\| \to 0} \frac{f(t + \epsilon) - f(t)}{\|\epsilon\|^{H'}} = 0\}. \tag{37}$$

3.2 Non-Gaussian Fields

In this section we will introduce stable self-similar fields. Through these examples we would like to recall that self-similar fields need not to be Gaussian or even to have finite expectation and variance. The interested reader can complete this quick introduction with [34].

3.2.1 Moving Average Fractional Stable Fields

To construct a moving average type of self-similar fields, let us recall the moving average representation of fractional Brownian fields (30) It will be the starting point of moving average fractional stable field (in short mafsf).

Definition 3.8. For $0 < H < 1$, and $H \neq \frac{d}{\alpha}$, a field $(X_H(t))_{t\mathbb{R}^d}$ is called a moving average fractional stable field if it admits

$$X_H(t) \stackrel{(d)}{=} \int_{\mathbb{R}^d} \left[\|t - s\|^{H-d/\alpha} - \|s\|^{H-d/\alpha} \right] M_\alpha(ds), \tag{38}$$

where M_α is a real symmetric random stable measure defined in Definition 2.2.

When we compare (30) and (38), we remark that the fractional power $H - d/2$ has been replaced by $H - d/\alpha$, which is consistent with the rule of the thumb that claims that Gaussian variables correspond to $\alpha = 2$.

Proposition 3.9. *For $0 < H < 1$, and $H \neq \frac{d}{\alpha}$, the moving average fractional stable motion is well defined by the formula (38) and it is a H-self-similar field with stationary increments.*

Proof. To show that the integral in (38) is well defined, we have to check that:

$$\int_{\mathbb{R}^d} \left| \|t - s\|^{H-d/\alpha} - \|s\|^{H-d/\alpha} \right|^\alpha ds < \infty. \tag{39}$$

The integral may diverge when $\|s\| \to \infty$, $\|s\| \to 0$ or $s \to t$. Let us rewrite the integral in (39)

$$\int_{S^d} \int_0^\infty \left| \|t - \rho u\|^{H-d/\alpha} - \|\rho u\|^{H-d/\alpha} \right|^\alpha \rho^{d-1} d\rho d\sigma(u), \tag{40}$$

where σ is the surface measure on the unit sphere S^d. When $\|s\| \to \infty$ or $\rho \to \infty$,

$$\left| \|t - \rho u\|^{H-d/\alpha} - \|\rho u\|^{H-d/\alpha} \right|^\alpha = |H - d/\alpha|^\alpha \rho^{H\alpha-d-\alpha} t.u + o(\rho^{H\alpha-d-\alpha}).$$

The integral in (40) is convergent when $\rho \to \infty$ since $H < 1$.

When $\|s\| \to 0$, if $H > d/\alpha$ the integrand is bounded, else

$$\left| \|t - \rho u\|^{H-d/\alpha} - \|\rho u\|^{H-d/\alpha} \right|^\alpha \sim \rho^{(H\alpha-d)}$$

and the integral in (40) is convergent when $\rho \to 0$ since $0 < H$.

The convergence when $s \to t$ is obtained as in the case $\|s\| \to 0$.

Let us check now the stationarity of the increments. Let $\theta = (\theta_1, \ldots, \theta_n)$ and $t = (t_1, \ldots, t_n) \in (\mathbb{R}^d)^n$, the logarithm of characteristic function of the increments of the mafsf is:

$$-\log \left(\mathbb{E} \exp \left(i \sum_{j=2}^n \theta_j (X_H(t_j) - X_H(t_1)) \right) \right)$$

$$= \int_{\mathbb{R}^d} \left| \sum_{j=2}^n \theta_j (\|t_j - s\|^{H-d/\alpha} - \|t_1 - s\|^{H-d/\alpha}) \right|^\alpha ds.$$

Hence, it is clear that for every $\delta \in \mathbb{R}^d$

$$(X_H(t_2) - X_H(t_1), \ldots, X_H(t_n) - X_H(t_1))$$

$$\overset{(d)}{=} (X_H(t_2 + \delta) - X_H(t_1 + \delta), \ldots, X_H(t_n + \delta) - X_H(t_1 + \delta)).$$

Since, for every $\epsilon > 0$,

$$-\log\left(\mathbb{E}\exp\left(i\sum_{j=1}^{n}\theta_j(X_H(\epsilon t_j))\right)\right)$$

$$= \int_{\mathbb{R}^d}\left|\sum_{j=1}^{n}\theta_j(\|\epsilon t_j - s\|^{H-d/\alpha} - \|s\|^{H-d/\alpha})\right|^{\alpha}ds.$$

By letting $s' = \epsilon s$, we get

$$-\log\left(\mathbb{E}\exp\left(i\sum_{j=1}^{n}\theta_j(\epsilon^H X_H(t_j))\right)\right) = -\log\left(\mathbb{E}\exp\left(i\sum_{j=1}^{n}\theta_j(X_H(\epsilon t_j))\right)\right),$$

which yields the self-similarity property. $\square$

3.2.2 Real Harmonizable Fractional Stable Fields

One can easily construct a stable counterpart of the harmonizable representation (32) of the fractional Brownian motion. In this article we focus on real valued harmonizable process. We recall that the limit (28)

$$\lim_{R\to+\infty}\int_{\mathbb{R}^d}f(\xi)M_{\alpha,R}(d\xi)$$

is our definition of $\int_{\mathbb{R}^d}f(\xi)M_{\alpha}(d\xi)$ for every function $f \in L^{\alpha}(\mathbb{R}^d)$ that satisfies (21). Up to a normalization constant, it is consistent with Theorem 6.3.1 in [34].

Then, one can define the real harmonizable fractional stable field (in short rhfsf).

Definition 3.10. A field $(X_H(x))_{x\mathbb{R}^d}$ is called real harmonizable fractional stable field if

$$X_H(x) \stackrel{(d)}{=} \int_{\mathbb{R}^d}\frac{e^{-ix.\xi}-1}{\|\xi\|^{H+d/\alpha}}M_{\alpha}(d\xi). \tag{41}$$

This last definition is the stable counterpart of the harmonizable representation of the fractional Brownian motion (32) where $H + 1/2$ has been replaced by $H + d/\alpha$, and where the normalizing constant C_H has been dropped. One can also prove the following proposition.

Proposition 3.11. *The real harmonizable fractional stable field is well defined by the formula (41) and it is a H-self-similar field with stationary increments.*

Proof. To show that the integral in (41) is well defined we have to check that:

$$\int_{\mathbb{R}^d} \left| \frac{e^{-ix.\xi} - 1}{\|\xi\|^{H+d/\alpha}} \right|^\alpha d\xi < \infty. \tag{42}$$

The integral may diverge when $\|\xi\| \to \infty$, $\|\xi\| \to 0$.
The integral in (42) is convergent when $\|\xi\| \to \infty$ since $H > 0$.
When $\|\xi\| \to 0$

$$\left| \frac{e^{-ix.\xi} - 1}{\|\xi\|^{H+d/\alpha}} \right|^\alpha \le C\|\xi\|^{\alpha - H\alpha - d}$$

and the integral in (42) is convergent when $\|\xi\| \to 0$ since $H < 1$.

Let us check now the stationarity of the increments. Let $\theta = (\theta_1, \dots, \theta_n)$ and $\mathbf{x} = (x_1, \dots, x_n) \in (\mathbb{R}^d)^n$, the logarithm of characteristic function of the increments of the rhfsf is:

$$- \log \left(\mathbb{E} \exp \left(i \sum_{j=2}^n \theta_j (X_H(x_j) - X_H(x_1)) \right) \right)$$

$$= \int_{\mathbb{R}^d} \left| \sum_{j=2}^n \theta_j \frac{e^{-ix_j.\xi} - e^{-ix_1.\xi}}{\|\xi\|^{H+d/\alpha}} \right|^\alpha d\xi$$

$$= \int_{\mathbb{R}^d} \left| \sum_{j=2}^n \theta_j \frac{e^{-ix_j.\xi} - 1}{\|\xi\|^{H+d/\alpha}} \right|^\alpha d\xi.$$

Hence, it is clear that for every $\delta \in \mathbb{R}^d$

$$(X_H(x_2) - X_H(x_1), \dots, X_H(x_n) - X_H(x_1))$$

$$\stackrel{(d)}{=} (X_H(x_2 + \delta) - X_H(x_1 + \delta), \dots, X_H(x_n + \delta) - X_H(x_1 + \delta)).$$

Since, for every $\epsilon > 0$,

$$- \log \left(\mathbb{E} \exp \left(i \sum_{j=1}^n \theta_j (X_H(\epsilon x_j)) \right) \right) = \int_{\mathbb{R}^d} \left| \sum_{j=1}^n \theta_j \frac{e^{-i\epsilon x_j.\xi} - 1}{\|\xi\|^{H+d/\alpha}} \right|^\alpha ds.$$

By letting $\xi' = \epsilon\xi$, we get

$$- \log \left(\mathbb{E} \exp \left(i \sum_{j=1}^n \theta_j (\epsilon^H X_H(x_j)) \right) \right) = - \log \left(\mathbb{E} \exp \left(i \sum_{j=1}^n \theta_j (X_H(\epsilon x_j)) \right) \right),$$

which yields the self-similarity property. $\qquad \square$

4 Lévy Fields

In this part we consider fractional fields that are non-Gaussian fields with finite variance. Moreover we will show that there are not self-similar and we will introduce the weaker property of asymptotic self-similarity that they enjoy. In many aspects these fields are intermediate between Gaussian and Stable fractional fields introduced before. Actually, it is well known that in some fields of applications data do not fit Gaussian models, see for instance [24, 36, 41] for image modeling. More recently other processes that are neither Gaussian nor stable have been proposed to model Internet traffic (cf. [10, 17]).

4.1 Moving Average Fractional Lévy Fields

First a counterpart of moving average fractional stable field is introduced, when a random Lévy measure of Sect. 2.2 is used in the integral representation.

Definition 4.1. For $0 < H < 1$, and $H \neq \frac{d}{2}$, let us call a real valued field $(X_H(t))_{t \in \mathbb{R}^d}$ which admits a well-balanced moving-average representation

$$X_H(t) \stackrel{(d)}{=} \int_{\mathbb{R}^d} \left(||t - s||^{H - \frac{d}{2}} - ||s||^{H - \frac{d}{2}} \right) M(ds),$$

where $M(ds)$ is a random Lévy measure defined by (5) that satisfies the finite moment assumption (4), a moving average fractional Lévy field (in short mafLf) with parameter H.

One can check that $s \mapsto ||t - s||^{H - \frac{d}{2}} - ||s||^{H - \frac{d}{2}}$ is square integrable with respect to ds the Lebesgue measure on $\mathbb{R}^d$. Hence X_H is well defined. For the sake of simplicity, we omit the case $d = 1$, $H = 1/2$: $X_{1/2}(t)$ is equal in distribution to $\int_0^t M(ds)$, which is a Lévy process.

Let us illustrate this construction with a simple example: $d = 1$ and $\mu(du) = \frac{1}{2}(\delta_{-1}(du) + \delta_1(du))$, where δ's are Dirac masses. In this case $M(ds)$ is a compound random Poisson measure and can be written as an infinite sum of random Dirac masses

$$M(ds) = \sum_{n \in \mathbb{Z}} \delta_{S_n}(ds)\varepsilon_n,$$

where $S_{n+1} - S_n$ are identically independent random variables with an exponential law, and ε_n are identically distributed independent Bernoulli random variables such that $\mathbf{P}(\varepsilon_n = 1) = \mathbf{P}(\varepsilon_n = -1) = 1/2$. The ε_n's are independent of the S_n's. Since the measure μ is finite and $\int_{\mathbb{R}} u\mu(du) = 0$, the corresponding mafLm is in this special case:

$$X(t) = \sum_{n=-\infty}^{+\infty} \varepsilon_n(|t - S_n|^{H-1/2} - |S_n|^{H-1/2}). \tag{43}$$

Even if the previous limit is in L^2 sense, it suggests that the regularity of the sample path can be governed by $H - 1/2$. In the following section we need other tools to prove this fact, but this guess happens to be true and it is quite different from the regularity of sample paths of fractional Brownian fields. Moreover one can have almost sure convergence by using shot noise series representation of the Lévy measure.

Because of the isometry property of the random Lévy measure, mafLf have finite second order moments, and moreover have the same covariance structure as the fractional Brownian field. But they have different distributions than the fractional Brownian field, in particular they are non-Gaussian.

Proposition 4.2. *The covariance structure of mafLfs is*

$$R(s,t) = \mathbb{E}(X(s)X(t)) = \frac{var(X(1))}{2}\{\|s\|^{2H} + \|t\|^{2H} - \|s - t\|^{2H}\}. \quad (44)$$

MafLfs have stationary increments.

Proof. The first claim is consequence of the isometry property for the random Lévy measure (8). Actually a consequence of (8) is that

$$\mathbb{E}\left(\int_{\mathbb{R}^d} f(s)M(ds) \int_{\mathbb{R}^d} g(s)M(ds)\right) = \int_{\mathbb{R}} u^2\mu(du) \int_{\mathbb{R}^d} f(s)g(s)\,ds. \quad (45)$$

Then $\mathbb{E}(X(s)X(t)) = \mathbb{E}B_H(s)B_H(t)$, since the fractional Brownian field and the mafLf have the same kernel in their integral representation. Finally (44) is a consequence of (29).

Let $\theta = (\theta_1, \ldots, \theta_n)$ and $t = (t_1, \ldots, t_n) \in (\mathbb{R}^d)^n$, the logarithm of characteristic function of the increments of the mafLf is:

$$-\log\left(\mathbb{E}\exp\left(i\sum_{j=2}^n \theta_j(X_H(t_j) - X_H(t_1))\right)\right)$$

$$= \int_{\mathbb{R}^d \times \mathbb{R}} \exp(i\sum_{j=2}^n u\theta_j(\|t_j - s\|^{H-d/2} - \|t_1 - s\|^{H-d/2}))$$

$$-1 - i\sum_{j=2}^n u\theta_j(\|t_j - s\|^{H-d/2} - \|t_1 - s\|^{H-d/2})\,ds\mu(du).$$

Then if we set $s' = t_1 + s$ and use the invariance by translation of the Lebesgue measure one gets

$$- \log \left(\mathbb{E} \exp \left(i \sum_{j=2}^{n} \theta_j (X_H(t_j) - X_H(t_1)) \right) \right)$$

$$\overset{(d)}{=} - \log \left(\mathbb{E} \exp \left(i \sum_{j=2}^{n} \theta_j X_H(t_j - t_1) \right) \right)$$

and the stationarity of the increments of the mafLf.   □

However mafLfs are not self-similar, but they enjoy an asymptotic property called asymptotic self-similarity at infinity, which is defined in the next proposition.

Proposition 4.3. *The mafLfs are asymptotically self-similar at infinity with parameter H*

$$\lim_{R \to +\infty} \left(\frac{X_H(Rt)}{R^H} \right)_{t \in \mathbb{R}^d} \overset{(d)}{=} \int_{\mathbb{R}} u^2 \mu(du) \times (B_H(t))_{t \in \mathbb{R}^d} , \tag{46}$$

where the convergence is the convergence of the finite dimensional margins and B_H is a fractional Brownian field of index H.

Proof. Let us consider the multivariate function:

$$g_{t,v,H}(R, s, u) = iu \sum_{k=1}^{n} v_k \frac{\|Rt_k - s\|^{H-d/2} - \|s\|^{H-d/2}}{R^H} \tag{47}$$

where $t = (t_1, \ldots, t_n) \in (\mathbb{R}^d)^n$, and $v = (v_1, \ldots, v_n) \in \mathbb{R}^n$. Then

$$\mathbb{E} \exp \left(i \sum_{k=1}^{n} v_k \frac{X_H(Rt_k)}{R^H} \right)$$

$$= \exp \left(\int_{\mathbb{R}^d \times \mathbb{R}} [\exp(g_{t,v,H}(R, s, u)) - 1 - g_{t,v,H}(R, s, u)] ds \mu(du) \right). \tag{48}$$

The change of variable $s = R\sigma$ is applied to the integral of the previous right hand term to get:

$$\int_{\mathbb{R}^d \times \mathbb{R}} [\exp(R^{-d/2} g_{t,v,H}(1, \sigma, u)) - 1 - R^{-d/2} g_{t,v,H}(1, \sigma, u)] R^d d\sigma \mu(du). \tag{49}$$

Then as $R \to +\infty$, a dominated convergence argument yields that

$$\lim_{R \to +\infty} \mathbb{E} \exp \left(i \sum_{k=1}^{n} v_k \frac{X_H(Rt_k)}{R^H} \right)$$

$$= \exp \left(\frac{1}{2} \int_{\mathbb{R}^d \times \mathbb{R}} g_{t,v,H}^2(1, \sigma, u) d\sigma \mu(du) \right). \tag{50}$$

Therefore the logarithm of the previous limit is

$$-\frac{1}{2} \int_0^{+\infty} u^2 \mu(du) \int_{\mathbb{R}^d} \left(\sum_{k=1}^n v_k (\|t_k - \sigma\|^{H-d/2} - \|\sigma\|^{H-d/2}) \right)^2 d\sigma, \quad (51)$$

and this last integral is the variance of $\sum_{k=1}^n v_k B_H(t_k)$ which concludes the proof of the convergence of finite dimensional margins. $\qquad\square$

4.1.1 Regularity of the Sample Paths

To investigate the regularity of the sample paths of mafLf, one can use the Kolmogorov theorem (see [9, 20] for proofs) to show that the sample paths are locally Hölder-continuous for every exponent $H' < H - d/2$, when $H > d/2$, in the sense of Definition 3.6. Let us recall the statement of Kolmogorov's theorem.

Theorem 4.4. *Let $(X_t, \ t \in [A, B]^d)$ be a random field. If there exist three positive constants α, β, C such that, for every $t, s \in [A, B]^d$*

$$\mathbb{E}|X_t - X_s|^\alpha \le C\|t - s\|^{d+\beta},$$

then, there exists a locally γ-Hölder continuous modification $\widetilde{X}$ of X for every $\gamma < \beta/\alpha$. It means that there exist a random variable $h(\omega)$, and a constant $\delta > 0$ such that

$$\mathbb{P}\left[\omega, \sup_{\|t-s\| \le h(\omega)} \frac{|\widetilde{X}_t(\omega) - \widetilde{X}_s(\omega)|}{\|t - s\|^\gamma} \le \delta \right] = 1.$$

The control needed to apply Kolmogorov-Chentsov's theorem are direct applications of the isometry property. The questions are then: What happens when $H < d/2$? If $H - d/2 > 0$, can we show that the "true" exponent is strictly larger than $H - d/2$? If we consider the integrand: $G(t, s) = \|t-s\|^{H-d/2} - \|s\|^{H-d/2}$, it is clear that, when $H - d/2 < 0$, $G(., s)$ is not locally bounded, and when $H > d/2$, it is not H'-Hölderian if $H' > H - d/2$ in a neighborhood of s. Following Rosinski's rule of the thumb in [31], it is known that the simple paths of the integral defining $X_H(t)$ cannot be "smoother" than the integrand G.

Let us now make precise statements.

Proposition 4.5. *If $H > d/2$, for every $H' < H - d/2$, there exists a continuous modification of the mafLf X_H such that such that almost surely the sample paths of X_H are locally H' Hölder continuous i.e.*

$$\mathbb{P}\left[\omega; \sup_{0<\|s-t\|<\epsilon(\omega), \|s\|\le 1, \|t\|\le 1} \left(\frac{X_H(s) - X_H(t)}{\|s - t\|^{H'}} \right) \le \delta \right] = 1 \quad (52)$$

where $\epsilon(\omega)$ is an almost surely positive random variable and $\delta > 0$. Moreover, for every $H' > H - d/2$, $\mathbb{P}(X_H \notin \mathscr{C}^{H'}) > 0$, where $\mathscr{C}^{H'}$ is the space of Hölder-continuous functions on $[0, 1]^d$. Furthermore, if the control measure μ of the random measure M is not finite, $\mathbb{P}(X_H \notin \mathscr{C}^{H'}) = 1$.

Proof. Because of the isometry property,

$$\mathbb{E}(X_H(s) - X_H(t))^2 = \mathbb{E}(B_H(s) - B_H(t))^2 = C\|t - s\|^{2H},$$

where B_H is a fractional Brownian field. The property (52) is then a direct consequence of Kolmogorov Theorem. To prove the second part of the proposition, Theorem 4 of [31] will be applied to X_H. First we take a separable modification of X_H with a separable representation. The next step is to use the symmetrization argument of Sect. 5 in [31] if μ is not already symmetric. Then we can remark that the kernel $t \to \|t - s\|^{H-d/2} - \|s\|^{H-d/2} \notin \mathscr{C}^{H'}$ for every $H' > H - d/2$, and the conclusion of Theorem 4 is applied to the measurable linear subspace $\mathscr{C}^{H'}$ to get $\mathbb{P}(X_H \notin \mathscr{C}^{H'}) > 0$. To show that this probability is actually one, we rely on a zero-one law. The process X_H can be viewed as an infinitely divisible law on the Banach space $\mathscr{C}[0, 1]$ of the continuous functions endowed with the supremum norm. Let us consider the map:

$$\varphi : \mathbb{R} \times \mathbb{R}^d \to \mathscr{C}[0, 1]$$

$$(u, s) \to u(\|. - s\|^{H-d/2} - \|s\|^{H-d/2}\|).$$

The random Lévy measure $F(df)$ of the infinitely divisible law defined by X_H is now given by $\varphi(\mu^{sym}(du) \times ds) = F(df)$, where μ^{sym} is the control measure of the symmetrized process. Hence $F((\mathscr{C}[0, 1] \setminus \mathscr{C}^{H'}) = +\infty$, if $\mu^{sym}(\mathbb{R}) = +\infty$. Corollary 11 of [16] and $\mathbb{P}(X_H \notin \mathscr{C}^{H'}) > 0$ yield the last result of the proposition. $\qquad \square$

Now let us go back to the case $H < d/2$.

Proposition 4.6. *If $H < d/2$, for every compact interval $K \subset \mathbb{R}^d$,*

$$\mathbb{P}(X_H \notin \mathscr{B}(K)) > 0,$$

where $\mathscr{B}(K)$ is the space of bounded functions on K.

In this case, we remark that $t \to \|t - s\|^{H-d/2} - \|s\|^{H-d/2} \notin \mathscr{B}(K)$ for every $s \in K$. The proposition is then proved by applying Theorem 4 of [31] to $\mathscr{B}(K)$.

4.1.2 Local Asymptotic Self-similarity

We will now investigate local self-similarity for mafLfs. Let us first recall the definition.

Definition 4.7. A field $(Y(x))_{x\in\mathbb{R}^d}$ is locally asymptotically self-similar (lass) at point x, if

$$\lim_{\varepsilon\to 0^+}\left(\frac{Y(x+\varepsilon u)-Y(x)}{\varepsilon^{h(x)}}\right)_{u\in\mathbb{R}^d}\overset{(d)}{=}(T_x(u))_{u\in\mathbb{R}^d},\qquad(53)$$

where the non-degenerate field $(T_x(u))_{u\in\mathbb{R}^d}$ is called the tangent field at point x of Y and the limit is in distribution for all finite dimensional margins of the fields. Furthermore, the field is lass with multifractional function h if for every $x\in\mathbb{R}^d$, it is lass at point x with index $h(x)$.

Let us make some comments about this definition. First, a non-degenerate field means that the tangent field is not the null function almost surely. It should be noted that mafLfs, in general, do not have a tangent field. In this section we focus on the truncated stable case. In view of Propositions 4.3 and 4.8, the truncated stable case can be viewed as a bridge between fBf and moving average fractional stable field. Let

$$\mu_{\alpha,1}(du)=\frac{\mathbf{1}_{\{|u|\le 1\}}du}{|u|^{1+\alpha}}$$

be a control measure in the sense of Sect. 2.2 associated to the Lévy random measure $M_{\alpha,1}$. Denote the corresponding mafLf by $X_{H,\alpha}$

$$X_{H,\alpha}(t)=\int_{\mathbb{R}^d}(\|t-s\|^{H-\frac{d}{2}}-\|s\|^{H-\frac{d}{2}})dM_{\alpha,1}(s).$$

Proposition 4.8. *Let us assume that $\widetilde{H}$ defined by $\widetilde{H}-\frac{d}{\alpha}=H-\frac{d}{2}$ is such that $0<\widetilde{H}<1$. The mafLf $X_{H,\alpha}$, with control measure*

$$\mu_{\alpha,1}(du)=\frac{\mathbf{1}_{\{|u|\le 1\}}du}{|u|^{1+\alpha}},$$

is locally self-similar with parameter $\widetilde{H}$. For every fixed $t\in\mathbb{R}^d$,

$$\lim_{\epsilon\to 0^+}\left(\frac{X_{H,\alpha}(t+\epsilon x)-X_{H,\alpha}(t)}{\epsilon^{\widetilde{H}}}\right)_{x\in\mathbb{R}^d}\overset{(d)}{=}(Y_{\widetilde{H}}(x))_{x\in\mathbb{R}^d},\qquad(54)$$

where the limit is in distribution for all finite dimensional margins of the field. The limit is a moving average fractional stable field that has a representation:

$$Y_{\widetilde{H}}(x)=\int_{\mathbb{R}^d}(\|x-\sigma\|^{\widetilde{H}-d/\alpha}-\|\sigma\|^{\widetilde{H}-d/\alpha})M_\alpha(d\sigma),\qquad(55)$$

where $M_\alpha(d\xi)$ is a stable α−symmetric random measure.

Proof. Since the mafLf has stationary increments we only have to prove the convergence for $t = 0$. We consider a multivariate function:

$$g_{t,v,H}(\epsilon, s, u) = iu \sum_{k=1}^{n} v_k \frac{\|\epsilon t_k - s\|^{H-d/2} - \|s\|^{H-d/2}}{\epsilon^{\widetilde{H}}}, \tag{56}$$

where $t \in (\mathbb{R}^d)^n$, and $v \in \mathbb{R}^n$. Then

$$\mathbb{E} \exp \left(i \sum_{k=1}^{n} v_k \frac{X_{\widetilde{H}}(\epsilon t_k)}{\epsilon^{\widetilde{H}}} \right)$$

$$= \exp \left(\int_{\mathbb{R}^d \times \mathbb{R}} [\exp(g_{t,v,H}(\epsilon, s, u)) - 1 - g_{t,v,H}(\epsilon, s, u)] ds \mu(du) \right). \tag{57}$$

Then the change of variable $\sigma = \frac{s}{\epsilon}$ is applied and $\widetilde{H}$ has been chosen such that the integral in the previous equation is now:

$$\int_{\mathbb{R}^d \times \mathbb{R}} [\exp(g_{t,v,H}(1, \sigma, \epsilon^{-d/\alpha} u)) - 1 - g_{t,v,H}(1, \sigma, \epsilon^{-d/\alpha} u)]$$

$$\mathbf{1}(|u| < 1) \epsilon^d d\sigma \frac{du}{|u|^{1+\alpha}}. \tag{58}$$

Let us set $w = \epsilon^{-d/\alpha} u$. The integral becomes

$$I(\epsilon) = \int_{\mathbb{R}^d \times \mathbb{R}} [\exp(g_{t,v,H}(1, \sigma, w)) - 1 - g_{t,v,H}(1, \sigma, w)]$$

$$\mathbf{1}(|w| < \epsilon^{-d/\alpha}) d\sigma \frac{dw}{|w|^{1+\alpha}}. \tag{59}$$

Let us recall that

$$- C(\alpha)|x|^\alpha = \int_{\mathbb{R}} [e^{ixr} - 1 - ixr\mathbf{1}(|r| \le \epsilon^{-d/\alpha})] \frac{dr}{|r|^{1+\alpha}}, \tag{60}$$

for every $\epsilon > 0$, where $C(\alpha) = 2 \int_0^{+\infty} (1 - \cos(r)) \frac{dr}{r^{1+\alpha}}$. Let us write

$$J_\epsilon(x) = \int_{\mathbb{R}} [e^{ixr} - 1 - ixr] \mathbf{1}(|r| \le \epsilon^{-d/\alpha}) \frac{dr}{|r|^{1+\alpha}}$$

then

$$\lim_{\epsilon \to 0^+} \left(J_{\epsilon(x)} + C(\alpha)|x|^\alpha \right) = \lim_{\epsilon \to 0^+} \int_{\mathbb{R}} [1 - e^{ixr}] \mathbf{1}(|r| > \epsilon^{-d/\alpha}) \frac{dr}{|r|^{1+\alpha}}$$

$$= 0.$$

Since $I(\epsilon) = \int_{\mathbb{R}^d} J_\epsilon(g_{t,v,H}(1,\sigma,1)) d\sigma$, a monotone convergence argument yields

$$\lim_{\epsilon \to 0^+} I(\epsilon) = -C(\alpha) \int_{\mathbb{R}^d} |g_{t,v,H}(1,\sigma,1)|^\alpha d\sigma. \tag{61}$$

Since this last expression is the logarithm of

$$\mathbb{E} \exp \left(i \sum_{k=1}^{n} v_k Y_{\widetilde{H}}(t_k) \right),$$

the proof is complete. $\qquad\square$

4.2 Real Harmonizable Fractional Lévy Fields

A counterpart of real harmonizable fractional stable field is introduced, when a random Lévy measure of Sect. 2.4, is used in the integral representation.

Definition 4.9. Let us call a real harmonizable fractional Lévy field (in short rhfLf), with parameter H, a real valued field $(X_H(t))_{t \in \mathbb{R}^d}$, which admits an harmonizable representation

$$X_H(x) \stackrel{(d)}{=} \int_{\mathbb{R}^d} \frac{e^{-ix\cdot\xi} - 1}{\|\xi\|^{\frac{d}{2}+H}} M(d\xi),$$

where $M(d\xi)$ is a complex isotropic random Lévy measure defined in Definition 2.5 that satisfies the finite moment assumption (16).

Please note that, since

$$\xi \mapsto \frac{e^{-ix\cdot\xi} - 1}{\|\xi\|^{\frac{d}{2}+H}}$$

satisfies (21) almost surely, rhfLfs are real valued! Because of the isometry property (26) of the complex isotropy random Lévy measure, rhfLfs have finite second order moments. As in the case of mafLf, they have the same covariance structure than fractional Brownian field s. But they have different distributions, in particular they are non-Gaussian. We will see later that the regularity properties of rhfLfs are quite different from those of mafLfs.

Proposition 4.10. *The covariance structure of rhfLfs is*

$$R(x, y) = \mathbb{E}(X(x)X(y)) = \frac{var(X(1))}{2}\{\|x\|^{2H} + \|y\|^{2H} - \|x - y\|^{2H}\}. \quad (62)$$

RhfLfs have stationary increments.

Proof. The first claim is consequence of the isometry property and of the fact that fractional Brownian field s and rhfLfs have the same kernel in their integral representation.

Let $\theta = (\theta_1, \ldots, \theta_n)$ and $\mathbf{x} = (x_1, \ldots, x_n) \in (\mathbb{R}^d)^n$, the logarithm of characteristic function of the increments of the rhfLf is:

$$-\log\left(\mathbb{E}\exp\left(i\sum_{j=2}^{n}\theta_j(X_H(x_j) - X_H(x_1))\right)\right)$$

$$= \int_{\mathbb{R}^d \times \mathbb{C}} \exp\left(i\sum_{j=2}^{n}\theta_j 2\Re\left(z\frac{e^{-ix_j\cdot\xi} - e^{-ix_1\cdot\xi}}{\|\xi\|^{H+d/2}}\right)\right)$$

$$-1 - i\sum_{j=2}^{n}\theta_j 2\Re\left(z\frac{e^{-ix_j\cdot\xi} - e^{-ix_1\cdot\xi}}{\|\xi\|^{H+d/2}}\right)\right)d\xi\mu(dz).$$

Then, if we set $z' = e^{-ix_1\cdot\xi}z$, and use the invariance by rotation of $\mu(dz)$ one gets

$$-\log\left(\mathbb{E}\exp\left(i\sum_{j=2}^{n}\theta_j(X_H(x_j) - X_H(x_1))\right)\right)$$

$$= -\log\left(\mathbb{E}\exp\left(i\sum_{j=2}^{n}\theta_j X_H(x_j - x_1)\right)\right),$$

and the stationarity of the increments of the rhfLf. $\qquad\square$

Now we investigate some properties of self-similarity and regularity types that the rhfLf shares with the fractional Brownian field. In the first part of this section, we prove that two asymptotic self-similarity properties are true for the rhfLf. In the second part, we see that almost surely the paths of the rhfLf are Hölder-continuous, with a pointwise Hölder exponent H. See Definitions 3.6 and 3.7.

4.2.1 Asymptotic Self-similarity

Since we know the characteristic function of stochastic integrals of the measure $M(d\xi)$, we can prove the local self-similarity of the rhfLfs. Actually it is a consequence of the homogeneity property of $\frac{1}{\|\xi\|^{\frac{d}{2}+H}}$ and of a central limit theorem for the stochastic measure $M(d\xi)$. In this case, one can achieve better convergence than the limit of the finite dimensional margins. In this case one speaks of strongly locally asymptotically self-similar fields.

Definition 4.11. A field $(Y(x))_{x \in \mathbb{R}^d}$ is strongly locally asymptotically self-similar (slass) at point x if

$$\lim_{\varepsilon \to 0+} \left(\frac{Y(x + \varepsilon u) - Y(x)}{\varepsilon^{h(x)}} \right)_{u \in \mathbb{R}^d} \stackrel{(d)}{=} (T_x(u))_{u \in \mathbb{R}^d}, \tag{63}$$

where the non-degenerate field $(T_x(u))_{u \in \mathbb{R}^d}$ is called the tangent field at point x of Y, and the limit is in distribution on the space of continuous functions endowed with the topology of the uniform convergence on every compact. Furthermore, the field is strongly lass with multifractional function h if for every $x \in \mathbb{R}^d$, it is strongly lass at point x with index $h(x)$.

When we compare the Definitions 4.7 and 4.11, the only difference is that the tightness is required in the later case.

Proposition 4.12. *The real harmonizable fractional Lévy field is strongly locally self similar with parameter H in the sense that for every fixed $x \subset \mathbb{R}^d$:*

$$\lim_{\epsilon \to 0+} \left(\frac{X_H(x + \epsilon u) - X_H(x)}{\epsilon^H} \right)_{u \in \mathbb{R}^d} \stackrel{(d)}{=}$$

$$C(H) \left(2\pi \int_0^{+\infty} \rho^2 \mu_\rho(d\rho) \right)^{1/2} (B_H(u))_{u \in \mathbb{R}^d}, \tag{64}$$

where B_H is a standard fractional Brownian field, and where the function C is given by

$$C(s) = \int_{\mathbb{R}^d} \frac{2(1 - \cos(\xi_1))}{\|\xi\|^{d+2s}} \frac{d\xi}{(2\pi)^{d/2}} \tag{65}$$

$$= \frac{\pi^{1/2} \Gamma(s + 1/2)}{2^{d/2} \Gamma(2s) \sin(\pi s) \Gamma(s + d/2)}. \tag{66}$$

A standard fractional Brownian field has a covariance given in (29) with $C = 1$.

Proof. The convergence of the finite dimensional margins is proved first. Since the rhfLf has stationary increments, we only have to prove the convergence for $x = 0$. Let us consider the multivariate function:

$$g_{u,v,H}(\epsilon, \xi, z) = i2\Re \left(z \sum_{k=1}^n v_k \frac{e^{-i\epsilon u_k \cdot \xi} - 1}{\epsilon^H \|\xi\|^{\frac{d}{2}+H}} \right) \tag{67}$$

where $u = (u_1, \ldots, u_n) \in (\mathbb{R}^n)^d$ and $v = (v_1, \ldots, v_n) \in \mathbb{R}^n$. Then

$$\mathbb{E} \exp \left(i \sum_{k=1}^{n} v_k \frac{X_H(\epsilon u_k)}{\epsilon^H} \right)$$
$$= \exp \left(\int_{\mathbb{R}^d \times \mathbb{C}} [\exp(g_{u,v,H}(\epsilon, \xi, z)) - 1 - g_{u,v,H}(\epsilon, \xi, z)] d\xi d\mu(z) \right). \quad (68)$$

The change of variable $\lambda = \epsilon \xi$ is applied to the integral of the previous right hand term to get

$$\int_{\mathbb{R}^d \times \mathbb{C}} [\exp(\epsilon^{d/2} g_{u,v,H}(1, \lambda, z)) - 1 - \epsilon^{d/2} g_{u,v,H}(1, \lambda, z)] \frac{d\lambda}{\epsilon^d} d\mu(z). \quad (69)$$

Then as $\epsilon \to 0^+$ a dominated convergence argument yields that

$$\lim_{\epsilon \to 0^+} \mathbb{E} \exp \left(i \sum_{k=1}^{n} v_k \frac{X_H(\epsilon u_k)}{\epsilon^H} \right)$$
$$= \exp \left(\frac{1}{2} \int_{\mathbb{R}^d \times \mathbb{C}} g_{u,v,H}^2(1, \lambda, z) d\lambda d\mu(z) \right). \quad (70)$$

Moreover (15) allows us to express the logarithm of the previous limit as:

$$-2\pi \int_0^{+\infty} \rho^2 \mu_\rho(d\rho) \int_{\mathbb{R}^d} \frac{|\sum_{k=1}^n v_k(e^{-iu_k \cdot \lambda} - 1)|^2}{\|\lambda\|^{d+2H}} d\lambda, \quad (71)$$

and this last integral is the variance of $C(H) \sum_{k=1}^n v_k B_H(u_k)$, which concludes the proof of the convergence of finite dimensional margins. $\qquad \square$

We refer to [9, 20] for a reminder on convergence in distribution in the space of continuous multivariate functions and recall the following theorem.

Theorem 4.13. *Let $(X^n)_{n \geq 1}$ and X be continuous random fields valued such that for all $k \in \mathbb{N}$ and for all $t_1, \ldots, t_k \in [A, B]^d$ the finite dimensional distributions of $(X^n(t_1), \ldots, X^n(t_k))$ converge to $(X(t_1), \ldots, X(t_k))$. If there exist three positive constants α, β and C such that, for $t, s \in [A, B]^d$,*

$$\sup_{n \geq 1} \mathbb{E}|X_t^n - X_s^n|^\alpha \leq C\|t - s\|^{d+\beta},$$

then X^n converges to the continuous field X in distribution on the space of continuous functions endowed with the topology of the uniform convergence.

Proof. Because of Theorem 4.13, we need to estimate

$$\mathbb{E}(X_H(x) - X_H(y))^{2p}$$

for p large enough. Unfortunately, when $H > 1 - d/2$, these moments are not finite because of the asymptotic of the integrand:

$$g_0(x, \xi) = \frac{e^{-ix\cdot\xi} - 1}{\|\xi\|^{\frac{d}{2}+H}} \tag{72}$$

when $\|\xi\| \to 0$. In the case $H > 1 - d/2$, we thus apply a transformation to the integrand g_0 to analyze in two different ways its behavior at both ends of the spectrum.

Let us first consider the easy case: $H \leq 1 - d/2$. Then $g_0(x, .) \in L^{2q}(\mathbb{R}^d)$ $\forall q \in \mathbb{N}^*$ and

$$\|g_0(x, .) - g_0(y, .)\|_{L^{2q}(\mathbb{R}^d)}^{2q} = \|x - y\|^{2Hq+d(q-1)} \|g_0(e_1, .)\|_{L^{2q}(\mathbb{R}^d)}^{2q}$$

where $e_1 = (1, 0, \ldots, 0) \in \mathbb{R}^d$. Because of (27) we know that

$$\mathbb{E}(X_H(x) - X_H(y))^{2p} = \sum_{n=1}^{p} D(n)\|x - y\|^{2Hp+d(p-n)}$$

for some nonnegative constants $D(n)$. Hence there exists $C < +\infty$ such that

$$\mathbb{E}(X_H(x) - X_H(y))^{2p} \leq C\|x - y\|^{2Hp} \tag{73}$$

where x, y are in a fixed compact. Hence if $H \leq 1 - d/2$

$$\mathbb{E}\left(\frac{(X_H(x + \epsilon u) - X_H(x + \epsilon v))^{2p}}{\epsilon^{2Hp}}\right) \leq \|u - v\|^{2Hp}$$

and one can take $p > \frac{d}{2H}$ to show the tightness .

When $H > 1 - d/2$, let us take K an integer such that $K \geq 1 + d/2$;

$$P_K(t) = \sum_{k=1}^{K} \frac{t^k}{k!},$$

and φ an even C^1-function such that $\varphi(t) = 1$ when $|t| \leq 1/2$ and $\varphi(t) = 0$ when $|t| > 1$. Then

$$g_K(x, \xi) = \frac{e^{-ix\cdot\xi} - 1 - P_K(-ix\cdot\xi)\varphi(\|x\|\|\xi\|)}{\|\xi\|^{\frac{d}{2}+H}}$$

is in $L^{2q}(\mathbb{R}^d)$ for every $x \in \mathbb{R}^d$ and $q \in \mathbb{N}^*$.

The field X_H is then split into two fields $X_H = X_H^+ + X_H^-$, where

$$X_H^+(x) = \int g_K(x, \xi) M(d\xi), \tag{74}$$

and

$$X_H^-(x) = \int \frac{P_K(-ix \cdot \xi)}{\|\xi\|^{\frac{d}{2}+H}} \varphi(\|x\| \|\xi\|) M(d\xi). \tag{75}$$

A method similar to the one used for X_H when $H \leq 1 - d/2$ is applied to X_H^+ and we check that X_H^- has almost surely C^1 paths.

Let us start by the remark that

$$\|g_K(x, .)\|_{L^{2q}(\mathbb{R}^d)}^{2q} = \|x\|^{2Hq+d(q-1)} \|g_K(e_1, .)\|_{L^{2q}(\mathbb{R}^d)}^{2q}.$$

As in the easy case we have to estimate when $\epsilon \to 0^+$

$$I_\epsilon = \int_{\mathbb{R}^d} |g_K(x, \xi) - g_K(x + \epsilon u, \xi)|^{2q} d\xi.$$

Let us split this integral into

$$I_\epsilon^+ = \int_{\epsilon \|\xi\| \geq 1} |g_K(x, \xi) - g_K(x + \epsilon u, \xi)|^{2q} d\xi,$$

and

$$I_\epsilon^- = \int_{\epsilon \|\xi\| < 1} |g_K(x, \xi) - g_K(x + \epsilon u, \xi)|^{2q} d\xi$$

as $I_\epsilon = I_\epsilon^+ + I_\epsilon^-$. Actually

$$|g_K(x, \xi) - g_K(y, \xi)| = |g_0(x - y, \xi)|$$

on $\{\epsilon \|\xi\| \geq 1\}$ for ϵ small enough, and we get by the change of variable $\lambda = \epsilon \xi$

$$I_\epsilon^+ = \epsilon^{2Hq+d(q-1)} \int_{\|\lambda\| \geq 1} \frac{|e^{-ie_1 \cdot \lambda} - 1|^{2q}}{\|\lambda\|^{2Hq+dq}} d\lambda. \tag{76}$$

Then a Taylor expansion is applied to I_ϵ^-

$$I_\epsilon^- = \int_{\|\xi\| < \frac{1}{\epsilon}} |dg_K(\theta(x, \epsilon u, \xi), \xi).\epsilon u|^{2q} d\xi,$$

where $dg_K(\theta(x, \epsilon u, \xi), \xi)$ is the differential of the map $g_K(., \xi)$ and $\theta(x, \epsilon, \xi)$ is a point in the segment $(x, x + \epsilon u)$. Note that

$$\int_{\|\xi\| < C} \|dg_K(\theta(x, \epsilon u, \xi), \xi)\|^{2q} d\xi < +\infty$$

for every fixed C, and that

$$\|dg_K(\theta(x, \epsilon u, \xi), \xi)\|^{2q} = O(\|\xi\|^{2q(1 - \frac{d}{2} - H)}) \quad \text{when} \quad \|\xi\| \to +\infty.$$

Hence

$$\left(\int_{\|\xi\| < \frac{1}{\epsilon}} \|dg_K(\theta(x, \epsilon u, \xi), \xi)\|^{2q} d\xi\right) \epsilon^{2q} = 0(\epsilon^{2Hq + d(q-1)}) \quad \text{when} \quad \epsilon \to 0^+$$

and

$$|I_\epsilon^-| \le C\epsilon^{2Hq + d(q-1)} \tag{77}$$

when $\epsilon \to 0^+$. Because of (76) and (77), there exists a positive constant C such that

$$\int_{\mathbb{R}^d} |g_K(x, \xi) - g_K(y, \xi)|^{2q} d\xi \le C\|x - y\|^{2Hq + d(q-1)},$$

and consequently

$$\mathbb{E}(X_H^+(x) - X_H^+(y))^{2p} \le C\|x - y\|^{2Hp}, \tag{78}$$

when $\|x\| \le 1$, $\|y\| \le 1$, which yields that the distributions of

$$\left(\frac{X_H^+(x + \epsilon.) - X_H^+(x)}{\epsilon^H}\right)_{\epsilon > 0}$$

are tight. To conclude, let us write X_H^- for $\|x\| < \epsilon$ as

$$\int_{\epsilon\|\xi\| \le 1/2} \frac{P_K(-ix \cdot \xi)}{\|\xi\|^{\frac{d}{2} + H}} M(d\xi) + \int_{1/2 \le \epsilon\|\xi\| \le 1} \frac{P_K(-ix \cdot \xi)}{\|\xi\|^{\frac{d}{2} + H}} \varphi(\|x\|\|\xi\|) M(d\xi).$$

The first integral of the previous line is actually a polynomial in the variables $(x_1, \ldots, x_d)$ with coefficients that are random variables, hence it has almost surely C^1 paths. Let us remark that the integrand of the second integral is bounded with compact support in $\mathbb{R}^d \times \mathbb{C}$ and is C^1 in the variable x, so is the integral which yields that $X_{\tilde{H}}^-$ is almost surely C^1. Then it is clear that

$$\lim_{\epsilon \to 0^+} \left(\frac{X_{\tilde{H}}^-(x + \epsilon u) - X_{\tilde{H}}^-(x)}{\epsilon^H} \right)_{u \in \mathbb{R}^d} \stackrel{(a.s.)}{=} 0$$

which concludes the proof.   □

We now want to exhibit an example of rhfLf that has asymptotic self-similarity properties when the increment is taken on large scales. Actually if the control measure $\mu_\rho(d\rho)$ is $\frac{d\rho}{|\rho|^{1+\alpha}}\mathbf{1}(|\rho| < 1)$ where $0 < \alpha < 2$, we show that at large scales the rhfLf is asymptotically self-similar with parameter $0 < \tilde{H} < 1$ such that $\tilde{H} + \frac{d}{\alpha} = H + \frac{d}{2}$.

Heuristically it means that at large scales the truncation of the Lévy measure disappears.

Moreover the limit field is a rhfsf with parameter $\tilde{H}$. This shows that at large scales the behavior of rhfLf can be very far from the Gaussian model even if the rhfLfs are fields that have moments of order 2. The rhfLf with control measure $\frac{d\rho}{|\rho|^{1+\alpha}}\mathbf{1}(|\rho| < 1)$ can be viewed roughly speaking as in between a rhfsf at large scales and a fractional Brownian field at low scales. Let us now state precisely the asymptotic self-similarity.

Proposition 4.14. *Let us assume that $\tilde{H}$ defined by $\tilde{H} + \frac{d}{\alpha} = H + \frac{d}{2}$ is such that $0 < \tilde{H} < 1$. The real harmonizable fractional Lévy field, with control measure $\mu_\rho(d\rho)$,*

$$\frac{d\rho}{|\rho|^{1+\alpha}}\mathbf{1}(|\rho| < 1),$$

is asymptotically self-similar at infinity with parameter $\tilde{H}$

$$\lim_{R \to +\infty} \left(\frac{X_H(Ru)}{R^{\tilde{H}}} \right)_{u \in \mathbb{R}^d} \stackrel{(d)}{=} (Y_{\tilde{H}}(u))_{u \in \mathbb{R}^d}, \tag{79}$$

where the limit is in distribution for all finite dimensional margins of the fields, and the limit is a real harmonizable fractional stable field that has a representation:

$$Y_{\tilde{H}}(u) = \int_{\mathbb{R}^d} \frac{e^{-iu \cdot \xi} - 1}{\|\xi\|^{\frac{d}{\alpha} + \tilde{H}}} M_\alpha(d\xi), \tag{80}$$

where $M_\alpha(d\xi)$ is complex isotropic α-stable random measure defined in (28).

Proof. As in Proposition 4.8, we consider a multivariate function:

$$g_{u,v,H}(R, \xi, z) = i2\Re \left(z \sum_{k=1}^n v_k \frac{e^{-iRu_k \cdot \xi} - 1}{R^{\tilde{H}} \|\xi\|^{\frac{d}{2} + H}} \right) \tag{81}$$

where $u \in (\mathbb{R}^n)^d$ and $v \in \mathbb{R}^n$. We write

$$\mathbb{E} \exp \left(i \sum_{k=1}^{n} v_k \frac{X_H(Ru_k)}{R^{\tilde{H}}} \right)$$
$$= \exp \left(\int_{\mathbb{R}^d \times \mathbb{C}} [\exp(g_{u,v,H}(R,\xi,z)) - 1 - g_{u,v,H}(R,\xi,z)] d\xi d\mu(z) \right). \quad (82)$$

Then the change of variable $\lambda = R\xi$ is applied and $\tilde{H}$ has been chosen such that integral in the previous equation is now

$$\int_{\mathbb{R}^d \times [0,2\pi] \times \mathbb{R}_*^+} [\exp(g_{u,v,H}(1,\lambda,R^{d/\alpha}\rho e^{i\theta})) - 1 - g_{u,v,H}(1,\lambda,R^{d/\alpha}\rho e^{i\theta})]$$
$$\mathbf{1}(|\rho| < 1)R^{-d}d\lambda d\theta \frac{d\rho}{|\rho|^{1+\alpha}}. \quad (83)$$

Let us set $r = R^{d/\alpha}\rho$ the integral becomes

$$I(R) = \int_{\mathbb{R}^d \times [0,2\pi] \times \mathbb{R}} [\exp(g_{u,v,H}(1,\lambda,re^{i\theta})) - 1 - g_{u,v,H}(1,\lambda,re^{i\theta})]$$
$$\mathbf{1}(|r| < R^{d/\alpha})d\lambda d\theta \frac{dr}{2|r|^{1+\alpha}}. \quad (84)$$

Let us recall (60)

$$-C(\alpha)|x|^\alpha = \int_{\mathbb{R}} [e^{ixr} - 1 - ixr\mathbf{1}(|r| \le R^{d/\alpha})] \frac{dr}{|r|^{1+\alpha}},$$

for every $R > 0$, where $C(\alpha) = \int_0^{+\infty}(1 - \cos(r))\frac{dr}{r^{1+\alpha}}$ is given in (10). If we write

$$J_R = \int_{\mathbb{R}} [e^{ixr} - 1 - ixr]\mathbf{1}(|r| \le R^{d/\alpha}) \frac{dr}{2|r|^{1+\alpha}},$$

then

$$\lim_{R \to +\infty} \left(J_R + \frac{C(\alpha)}{2}|x|^\alpha \right) = \lim_{R \to +\infty} \int_{\mathbb{R}} [e^{ixr} - 1]\mathbf{1}(|r| > R^{d/\alpha}) \frac{dr}{2|r|^{1+\alpha}}$$
$$= 0.$$

Please remark that

$$\int_{\mathbb{R}} [e^{rg_{u,v,H}(1,\lambda,re^{i\theta})} - 1]\mathbf{1}(|r| > R^{d/\alpha})\frac{dr}{2|r|^{1+\alpha}}$$
$$= \int_{R^{d/\alpha}}^{\infty} (\cos(|g_{u,v,H}(1,\lambda,re^{i\theta})|) - 1)\frac{dr}{|r|^{1+\alpha}},$$

and hence is a non positive function. Moreover it is increasing with respects to R, and

$$\int_{\mathbb{R}^d \times [0,2\pi]} \int_{R^{d/\alpha}}^{\infty} [\cos(|g_{u,v,H}(1,\lambda,re^{i\theta})|) - 1]\frac{dr}{|r|^{1+\alpha}} d\lambda d\theta < \infty.$$

Hence, by monotone convergence,

$$\lim_{R \to +\infty} I(R) = -\frac{C(\alpha)}{2} \int_{\mathbb{R}^d \times [0,2\pi]} \left| 2\Re\left(e^{i\theta} \sum_{k=1}^{n} v_k \frac{e^{-iu_k\lambda} - 1}{\|\lambda\|^{\frac{d}{\alpha}+\tilde{H}}} \right) \right|^{\alpha} d\lambda d\theta, \quad (85)$$

which is also

$$-\frac{C(\alpha))}{2} \int_0^{2\pi} |2\cos(\theta)|^{\alpha} d\theta \int_{\mathbb{R}^d} \left| \sum_{k=1}^{n} v_k \frac{e^{-iu_k\lambda} - 1}{\|\lambda\|^{\frac{d}{\alpha}+\tilde{H}}} \right|^{\alpha} d\lambda. \quad (86)$$

Since this last expression is the logarithm of

$$\mathbb{E} \exp\left(i \sum_{k=1}^{n} v_k Y_{\tilde{H}}(u_k) \right),$$

(cf. (28)), the proof is complete.   □

4.2.2 Regularity of the Sample Paths of the rhfLf

We will show using the Theorem 4.4 that H is the pointwise Hölder exponent of the sample paths of the rhfLfs. Let us recall the definition of the pointwise exponent $H_f(x)$ of a deterministic function f at point x by

$$H_f(x) = \sup\{H', \quad \lim_{\epsilon \to 0} \frac{f(x+\epsilon) - f(x)}{\|\epsilon\|^{H'}} = 0\}. \quad (87)$$

Then the regularity of the sample paths is described by the following proposition.

Proposition 4.15. *For every $H' < H$, there exists a continuous modification of the rhfLf X_H such that almost surely the sample paths of X_H are locally H' Hölder continuous i.e.*

$$\mathbb{P}\left[\omega; \sup_{0<\|x-y\|<\epsilon(\omega),\|x\|\leq 1,\|y\|\leq 1}\left(\frac{X_H(x)-X_H(y)}{\|x-y\|^{H'}}\right)\leq\delta\right]=1, \qquad (88)$$

where $\epsilon(\omega)$ is an almost surely positive random variable and $\delta > 0$. Moreover, at every point x, the pointwise exponent $H_{X_H}(x)$ of the rhfLf X_H is almost surely equal to H.

Proof. In the first part of the proof, we will use the estimation of the moments

$$\mathbb{E}(X_H(x)-X_H(y))^{2p}$$

performed in the proof of Proposition 4.12 and Kolmogorov Theorem. When $H \leq 1 - d/2$, we already know by (73) that

$$\mathbb{E}(X_H(x)-X_H(y))^{2p} \leq C\|x-y\|^{2Hp},$$

when $\|x\| < 1$, $\|y\| < 1$ and Kolmogorov Theorem yields (88) for every $H' < H$. When $H > 1 - d/2$, we recall that X_H has been split into

$$X_H = X_H^+ + X_H^-,$$

where X_H^+ and X_H^- are defined in (74) and (75). Furthermore we know that X_H^+ is H'-Hölder-continuous for every $H' < H$ by Kolmogorov Theorem and the inequality (78), and that X_H^- has almost surely C^1 sample paths, which concludes the proof of (88).

Because of (52), at every point x the Hölder exponent satisfies $H(x) \geq H$. To show $H(x) \leq H$ let us use the local self-similarity (64). Actually, if $H' > H$, we can deduce from (64) that

$$\lim_{\epsilon\to 0^+}\frac{\epsilon^{H'}}{|X_H(x+\epsilon)-X_H(x)|} \overset{(d)}{=} 0,$$

which is also a convergence in probability. Hence we can find a sequence $(\epsilon_n)_{n\in\mathbb{N}} \to 0^+$ such that:

$$\lim_{n\to +\infty}\frac{|X_H(x+\epsilon_n)-X_H(x)|}{\epsilon_n^{H'}} = +\infty \quad \text{almost surely.}$$

This argument concludes the proof of Proposition 4.5. □

4.3   A Comparison of Lévy Fields

In the previous sections, we have introduced two different models of fractional fields with finite second moments that have the same covariance structure as the fractional Brownian field. Since both models have very different sample paths properties, it shows that the covariance structure cannot characterize the behavior of the smoothness of fractional fields. For modelling purpose, a comparison of mafLf and rhfLf may be useful.

First we can remark that the kernel of the rhfLf is the Fourier transform of the kernel of the mafLf i.e.

$$\widehat{\frac{e^{-it\xi} - 1}{C_H^{1/2}\|\xi\|^{H+d/2}}}(s) = D(H) \left(\|t - s\|^{H-d/2} - \|s\|^{H-d/2} \right), \qquad (89)$$

where $D(H)$ is a known function of H (cf. [9]). In the case of fractional Lévy fields, the Fourier transform of a Lévy measure is not a Lévy measure and it explains why rhfLf and mafLf have different distributions. Nevertheless, the self-similarity of fractional Lévy fields is reminiscent of this fact. Indeed in the limit in Proposition 4.3 for mafLf and in Proposition 4.12 for rhfLf is a fractional Brownian field. But in Proposition 4.3 the large scale behavior of mafLf is investigated, whereas Proposition 4.12 is concerned with the small scale behavior of rhfLf. It reminds us of the fact that the large scale behavior of a Fourier transform of a function is related to the small scale behavior of this function. The same remark is true, when we compare Proposition 4.8 and Proposition 4.14. Even if the limits are not the same in distribution in this last case they are of the same stable type.

4.4   Real Harmonizable Multifractional Lévy Fields

In this section, we introduce a class of Lévy fields proposed in [21] that yields multifractional fields related to rhfLf. The main idea behind multifractional fields in general is the fact that for some applications models like rhfLf or fBf parameterized by an index $0 < H < 1$ may not be rich enough. In some instance the practitioners may want models that behave locally, roughly speaking, as fractional models, but where the index will depend on the location. In multifractional models, the parameter H becomes a function h. A classical model for Gaussian fields is the multifractional Brownian field. (See [6, 9].)

Starting from the harmonizable representation of rhfLf one can propose fields called real harmonizable multifractional Lévy fields that are non Gaussian counterpart to multifractional Brownian fields.

Definition 4.16. Let $h : \mathbb{R}^d \rightarrow (0, 1)$ be a measurable function. A real valued field is called a real harmonizable multifractional Lévy field (in short rhmLf) with multifractional function h, if it admits the harmonizable representation

$$X_h(x) \stackrel{(d)}{=} \frac{1}{(C(h(x)))^{1/2}} \int_{\mathbb{R}^d} \frac{e^{-ix.\xi} - 1}{\|\xi\|^{d/2+h(x)}} M(d\xi), \tag{90}$$

where M is a complex isotropic random Lévy measure, and the normalization function C is given by (65).

Remark 4.17. If the multifractional function $h(x) = H$ is a constant then the rhmLf is a rhfLf. The multiplicative deterministic function in (90) is such that

$$\mathbb{E}X_h^2(x) = 1$$

for every $x \in S^{d-1}$.

In the following we list some properties of rhmLf referring to [21] for the proofs, and we suppose that the multifractional function h is β-Hölder continuous.

Proposition 4.18. *Let h be a function $h : \mathbb{R}^d \mapsto (0,1)$, and X_h be the corresponding rhmLf. Let K be a compact set, and $m = \inf\{h(x), \ x \in K\}$. For every $H < \min(m, \beta)$ there exists a modification of X_h such that almost surely the sample paths of X_h are H-Hölder continuous on K.*

RhmLfs are strongly locally asymptotically self-similar.

Proposition 4.19. *Let $h : \mathbb{R}^d \mapsto (0,1)$ be a β-Hölder continuous multifractional function, and X_h the corresponding rhmLf. Let us assume $\beta > \sup_{x\in\mathbb{R}^d} h(x)$, then, for every $x \in \mathbb{R}^d$, the rhmLf is strongly locally asymptotically self-similar, with tangent field a fractional Brownian field with Hurst exponent $H = h(x)$. More precisely*

$$\lim_{\varepsilon\to 0^+} \left(\frac{X_h(x+\varepsilon u) - X_h(x)}{\varepsilon^{h(x)}}\right)_{u\in\mathbb{R}^d} \stackrel{(d)}{=} \sqrt{4\pi \int_0^{+\infty} \rho^2 \mu_\rho(d\rho)} (B_H(u))_{u\in\mathbb{R}^d}, \tag{91}$$

where $H = h(x)$ and B_H is a fractional Brownian field with Hurst exponent H.

One can show the following result.

Corollary 4.20. *Let h be function $h : \mathbb{R}^d \mapsto (0,1)$ locally Hölder with exponent β. Let X_h be the corresponding rhmLf and fix x such that $h(x) < \beta$, the pointwise Hölder exponent of X_h at x is*

$$\sup\{H', \ \lim_{\epsilon\to 0} \frac{X_h(x+\epsilon) - X_h(x)}{|\epsilon|^{H'}} = 0\} = h(x) \tag{92}$$

almost surely.

Proposition 4.19 and Corollary 4.20 show that rhmLfs enjoy most of the properties of rhfLfs with a varying index given by the multifractional function h.

5   Statistics

In this section, we would like to discuss the use of the models introduced previously. One of the major question is the estimation of the various parameters in those models. The common framework of this estimation is that we observe only one sample path of the field on a finite set of locations in a compact set. Most of the results will then be asymptotic, when the mesh of the grid of the locations, where the fields are observed, is decreasing to 0. In short we are doing fill-in statistics. We will focus on rhfLfs and mafLfs.

5.1   Estimation for Real Harmonizable Fractional Lévy Fields

Let M a complex isotropic Lévy measure introduced in Definition 2.5 satisfying the assumptions (15) and (16). Let X_H be a rhfLf with a representation

$$X_H(x) = \int_{\mathbb{R}^d} \frac{e^{-ix \cdot \xi} - 1}{\|\xi\|^{\frac{d}{2}+H}} M(d\xi)$$

and for $\mathbf{k} = (k_1, \ldots, k_d) \in \mathbf{N}^d$ and $n \in \mathbf{N}^\star$, let us define $\frac{\mathbf{k}}{n} = \left(\frac{k_1}{n}, \ldots, \frac{k_d}{n}\right)$ and $X_H\left(\frac{\mathbf{k}}{n}\right) = X_H\left(\frac{k_1}{n}, \ldots, \frac{k_d}{n}\right)$.

The aim of this section is to perform the identification of the fractional index H in a semi-parametric setup from discrete observations of the field X_H on $[0, 1]^d$: the control measure $\mu(dz)$ of M is therefore unknown. The field X_H is observed at $\left(\frac{k_1}{n}, \ldots, \frac{k_d}{n}\right)$, $0 \leq k_i \leq n$, $i = 1, \ldots, d$.

Let $(a_\ell), \ell = 0, \ldots, K$ be a real valued sequence such that

$$\sum_{\ell=0}^{K} a_\ell = 0, \quad \sum_{\ell=0}^{K} \ell a_\ell = 0. \tag{93}$$

For $\mathbf{k} = (k_1, \ldots, k_d) \in \mathbf{N}^d$, define $a_\mathbf{k} = a_{k_1} \ldots a_{k_d}$. One can take for instance $K = 2$, $a_0 = 1$, $a_1 = -2$, $a_2 = 1$.

Define the increments of field X_H associated with the sequence a

$$\Delta X_\mathbf{p} = \sum_{\mathbf{k}=0}^{\mathbf{K}} a_\mathbf{k} X_H\left(\frac{\mathbf{k}+\mathbf{p}}{n}\right)$$

$$= \sum_{k_1,\ldots,k_d=0}^{K} a_{k_1} \ldots a_{k_d} X_H\left(\frac{k_1 + p_1}{n}, \ldots, \frac{k_d + p_d}{n}\right).$$

Define the quadratic variations associated with sequence a

$$Q_n = \frac{1}{(n-K+1)^d} \sum_{\mathbf{p}=0}^{\mathbf{n-K}} (\varDelta X_{\mathbf{p}})^2.$$

One can check that

$$\log(\mathbf{E}(Q_n)) = -2H \log n + C,$$

where C is a constant, and it is usual to identify H as the slope of a linear regression of $\log(Q_n)$ with respects of $\log n$ for the fractional Brownian motion. We refer to [8] to show that the quadratic variations are optimal in a Gaussian framework. For the sake of simplicity, we consider that the estimator of fractional index H is

$$\widehat{H}_n = \frac{1}{2} \log_2 \frac{Q_{n/2}}{Q_n},$$

but linear regression with $(\log(Q_{n/l}))_{l=1,..,L}$ could have been chosen, which should yield more robust results.

Actually this estimator is the estimator used by [15] to estimate the fractional index of a Fractional Brownian Motion.

Other estimators using wavelets coefficients instead of discrete variations are also available in the literature cf [4, 14] but only in the Gaussian and stable frameworks.

Theorem 5.1. *As $n \to +\infty$, one has:*

$$\widehat{H}_n \overset{(\mathbf{P})}{\to} H,$$

where $\overset{(\mathbf{P})}{\to}$ means a convergence in probability. Moreover $\forall \epsilon > 0, \; \exists M > 0$ such that

$$\sup_{N \in \mathbb{N}} \mathbf{P}(|\widehat{H}_N - H| > M N^{-\frac{1}{2}})) < \epsilon.$$

Proof. First define the following constants A and B:

$$A = 4\pi \int_{\mathbb{R}+} \rho^2 \mu_\rho(d\rho)$$

$$B = 4\pi \int_{\mathbb{R}+} \rho^4 \mu_\rho(d\rho).$$

Define the following two functional spaces:

$$\mathscr{F}_2 = \{f \in L^2(\mathbb{R}^d), \; f(-\xi) = \overline{f(\xi)}, \forall \xi \in \mathbb{R}^d\},$$
$$\mathscr{F}_4 = \{f \in L^2(\mathbb{R}^d) \cap L^4(\mathbb{R}^d), \; f(-\xi) = \overline{f(\xi)}, \forall \xi \in \mathbb{R}^d\}.$$

According to Proposition 2.6, we then have

- For all $f_1, f_2 \in \mathscr{F}_2$,

$$\mathbb{E} \int f_1(\xi) M(d\xi) \int f_2(\xi) M(d\xi) = A \int f_1(\xi) f_2(-\xi) d\xi. \qquad (94)$$

- For all $f_1, f_2, f_3, f_4 \in \mathscr{F}_4$,

$$\mathbb{E} \prod_{i=1}^{4} \int f_i(\xi) M(d\xi) = A^2 \left(\int f_1(\xi) f_2(-\xi) d\xi \times \int f_3(\xi) f_4(-\xi) d\xi \right.$$

$$+ \int f_1(\xi) f_3(-\xi) d\xi \times \int f_2(\xi) f_4(-\xi) d\xi$$

$$+ \left. \int f_1(\xi) f_4(-\xi) d\xi \times \int f_2(\xi) f_3(-\xi) d\xi \right)$$

$$+ B \left(\int f_1(\xi) f_2(-\xi) f_3(\xi) f_4(-\xi) d\xi \right.$$

$$+ \int f_1(\xi) f_2(\xi) f_3(-\xi) f_4(-\xi) d\xi$$

$$+ \left. \int f_1(\xi) f_2(-\xi) f_3(-\xi) f_4(\xi) d\xi \right). \qquad (95)$$

Define now $V_n = n^{2H} Q_n$.
Expectation of V_n
We deduce from (94)

$$\mathbb{E}(\Delta X_{\mathbf{p}})^2 = A \int_{\mathbb{R}^d} \frac{\left| \sum_{k=0}^{K} a_{\mathbf{k}} e^{i \frac{\mathbf{k}}{n} \cdot \xi} \right|^2}{||\xi||^{d+2H}} d\xi.$$

The change of variables $\lambda = \dfrac{\xi}{n}$ leads to

$$\mathbb{E}(\Delta X_{\mathbf{p}})^2 = A n^{-2H} \int_{\mathbb{R}^d} \frac{\left| \sum_{k=0}^{K} a_{\mathbf{k}} e^{i \mathbf{k} \cdot \lambda} \right|^2}{||\lambda||^{d+2H}} d\lambda,$$

and therefore

$$\mathbb{E} V_n = A \int_{\mathbb{R}^d} \frac{\left| \sum_{k=0}^{K} a_{\mathbf{k}} e^{i \mathbf{k} \cdot \lambda} \right|^2}{||\lambda||^{d+2H}} d\lambda.$$

Variance of V_n

We deduce from (95):

$$\mathbb{E}\left[(\Delta X_{\mathbf{p}})^2(\Delta X_{\mathbf{p}'})^2\right] = T_1 + T_2 + T_3 + T_4,$$

with

$$T_1 = A^2 \left(\int_{\mathbb{R}^d} \frac{\left| \sum_{k=0}^{K} a_k e^{-i\frac{k}{n}\cdot\xi} \right|^2}{||\xi||^{d+2H}} d\xi \right)^2,$$

$$T_2 = 2A^2 \left(\int_{\mathbb{R}^d} e^{-i\frac{\mathbf{p}-\mathbf{p}'}{n}\cdot\xi} \frac{\left| \sum_{k=0}^{K} a_k e^{-i\frac{k}{n}\cdot\xi} \right|^2}{||\xi||^{d+2H}} d\xi \right)^2,$$

$$T_3 = 2B \int_{\mathbb{R}^d} \frac{\left| \sum_{k=0}^{K} a_k e^{i\frac{k}{n}\cdot\xi} \right|^4}{||\xi||^{2d+4H}} d\xi,$$

$$T_4 = B \int_{\mathbb{R}^d} e^{-i\frac{2\mathbf{p}-2\mathbf{p}'}{n}\cdot\xi} \frac{\left| \sum_{k=0}^{K} a_k e^{i\frac{k}{n}\cdot\xi} \right|^4}{||\xi||^{2d+4H}} d\xi.$$

So that

$$\mathrm{var}V_n = G_n + N_n, \tag{96}$$

with

$$G_n = 2A^2 \frac{n^{4H}}{(n-K+1)^{2d}} \sum_{\mathbf{p},\mathbf{p}'=0}^{\mathbf{n}-\mathbf{K}} \left(\int_{\mathbb{R}^d} e^{-i\frac{\mathbf{p}-\mathbf{p}'}{n}\cdot\xi} \frac{\left| \sum_{k=0}^{K} a_k e^{i\frac{k}{n}\cdot\xi} \right|^2}{||\xi||^{d+2H}} d\xi \right)^2,$$

$$N_n = 2Bn^{4H} \int_{\mathbb{R}^d} \frac{\left| \sum_{k=0}^{K} a_k e^{i\frac{k}{n}\cdot\xi} \right|^4}{||\xi||^{2d+4H}} d\xi \tag{97}$$

$$+B\frac{n^{4H}}{(n-K+1)^{2d}} \sum_{\mathbf{p},\mathbf{p}'=0}^{\mathbf{n}-\mathbf{K}} \int_{\mathbb{R}^d} e^{-i\frac{2\mathbf{p}-2\mathbf{p}'}{n}\cdot\xi} \frac{\left| \sum_{k=0}^{K} a_k e^{i\frac{k}{n}\cdot\xi} \right|^4}{||\xi||^{2d+4H}} d\xi \tag{98}$$

We first study the part G_n of the variance.

The change of variables $\lambda = \dfrac{\xi}{n}$ leads to

$$
\int_{\mathbb{R}^d} e^{-i\frac{\mathbf{p}-\mathbf{p}'}{n}\cdot\xi} \frac{\left|\sum_{\mathbf{k}=0}^{\mathbf{K}} a_{\mathbf{k}} e^{-i\frac{\mathbf{k}}{n}\cdot\xi}\right|^2}{||\xi||^{d+2H}} d\xi = n^{-2H} \int_{\mathbb{R}^d} e^{-i(\mathbf{p}-\mathbf{p}')\cdot\lambda} \frac{\left|\sum_{\mathbf{k}=0}^{\mathbf{K}} a_{\mathbf{k}} e^{i\mathbf{k}\cdot\lambda}\right|^2}{||\lambda||^{d+2H}} d\lambda.
$$
(99)

Define the operator $\mathbf{D} = \displaystyle\prod_{j=1}^{d} \frac{\partial}{\partial x_i}$. Let us suppose that $\forall j, \quad p_j \neq p'_j$, integrating by parts leads to

$$
\int_{\mathbb{R}^d} e^{-i(\mathbf{p}-\mathbf{p}')\cdot\lambda} \frac{\left|\sum_{\mathbf{k}=0}^{\mathbf{K}} a_{\mathbf{k}} e^{i\mathbf{k}\cdot\lambda}\right|^2}{||\lambda||^{d+2H}} d\lambda
$$

$$
= i^d \prod_{j=1}^{d} \frac{1}{(p_j - p'_j)} \int_{\mathbb{R}^d} e^{-i(\mathbf{p}-\mathbf{p}')\cdot\lambda} \mathbf{D}\left[\frac{\left|\sum_{\mathbf{k}=0}^{\mathbf{K}} a_{\mathbf{k}} e^{i\mathbf{k}\cdot\lambda}\right|^2}{||\lambda||^{d+2H}}\right] d\lambda.
$$
(100)

The conditions $\displaystyle\sum_{\ell=0}^{K} a_\ell = 0, \quad \sum_{\ell=0}^{K} \ell a_\ell = 0$ ensure the convergence of the integral.

Since there exists a constant C_1 such that, as $n \to +\infty$,

$$
\left(\frac{1}{n} \sum_{m,m'=0, m \neq m'}^{n-K} \frac{1}{(m-m')^2}\right) \to C_1,
$$

$$
n^d G_n \to C_2.
$$
(101)

We know study the part N_n of the variance. Using the change of variables $\lambda = \dfrac{\xi}{n}$, one obtains for T_3

$$
\frac{2B}{n^d} \int_{\mathbb{R}^d} \frac{\left|\sum_{\mathbf{k}=0}^{\mathbf{K}} a_{\mathbf{k}} e^{i\mathbf{k}\cdot\lambda}\right|^4}{||\lambda||^{2d+4H}} d\lambda.
$$

It remains to study the part of N_n depending of T_4. The line (98) can be written as

$$
\frac{B}{n^d} \frac{1}{(n-K+1)^{2d}} \sum_{\mathbf{p},\mathbf{p}'=0}^{n-K} \int_{\mathbb{R}^d} e^{2i(\mathbf{p}-\mathbf{p}')\cdot\lambda} \frac{\left|\sum_{\mathbf{k}=0}^{\mathbf{K}} a_{\mathbf{k}} e^{i\mathbf{k}\cdot\lambda}\right|^4}{||\lambda||^{2d+4H}} d\lambda .
$$

Using as previously integration by parts, one proves that (98) is negligible with respect to (97). Hence

$$n^d N_n \to C_3. \tag{102}$$

To sum up, because of (96), (101), and (102), we have proved that there exists $C > 0$ such that

$$n^d \mathrm{var} V_n \to C,$$

and Theorem 5.1 is proved. $\square$

5.2 Identification of mafLf

We now perform the identification for mafLfs with truncated stable control measures. The random Lévy measure M is therefore associated with $\mu(du) = \frac{1_{\{|u| \leq 1\}} du}{|u|^{1+\alpha}}$. Let us recall that the corresponding mafLf is denoted by

$$X_{H,\alpha}(t) = \int_{\mathbb{R}^d} \|t - s\|^{H-d/2} - \|s\|^{H-d/2} dM(s).$$

As in the previous section, for $\mathbf{k} = (k_1, \ldots, k_d) \in \mathbf{N}^d$ and $n \in \mathbf{N}^\star$, let us define

$$\frac{\mathbf{k}}{2^n} = \left(\frac{k_1}{2^n}, \ldots, \frac{k_d}{2^n} \right),$$

$$X_{H,\alpha}\left(\frac{\mathbf{k}}{2^n} \right) = X_{H,\alpha}\left(\frac{k_1}{2^n}, \ldots, \frac{k_d}{2^n} \right).$$

The aim of this section is to perform the identification of the fractional indexes H and $\widetilde{H} = H - \frac{d}{2} + \frac{d}{\alpha}$, or equivalently of indexes H and α, with discrete observations of field $X_{H,\alpha}$ on $[0,1]^d$. The field $X_{H,\alpha}$ is observed at times $\left(\frac{k_1}{2^n}, \ldots, \frac{k_d}{2^n} \right)$, $0 \leq k_i \leq 2^n$, $i = 1, \ldots, d$.

Let (a_ℓ), $\ell = 0, \ldots, K$ be a real valued sequence such that:

$$\sum_{\ell=0}^{K} a_\ell = 0, \quad \sum_{\ell=0}^{K} \ell a_\ell = 0. \tag{103}$$

From now on, multi-indexes are written with bold letters. For $\mathbf{k} = (k_1, \ldots, k_d) \in \mathbf{N}^d$ define:

$$a_{\mathbf{k}} = a_{k_1} \ldots a_{k_d}.$$

Define the increments of field $X_{H,\alpha}$ associated with the sequence a:

$$\Delta X_{\mathbf{p},n} = \sum_{\mathbf{k}=0}^{\mathbf{K}} a_{\mathbf{k}} X_{H,\alpha} \left(\frac{\mathbf{k}+\mathbf{p}}{2^n} \right)$$

$$\stackrel{def}{=} \sum_{k_1,\ldots,k_d=0}^{K} a_{k_1} \ldots a_{k_d} X_{H,\alpha} \left(\frac{k_1+p_1}{2^n}, \ldots, \frac{k_d+p_d}{2^n} \right),$$

one can for instance take $K = 2$, $a_0 = 1$, $a_1 = -2$, $a_2 = 1$.

For $\beta > 0$, define the β-variations by

$$V_{n,\beta} = \frac{1}{(2^n - K)^d} \sum_{\mathbf{p}=1}^{2^n-\mathbf{K}} |\Delta X_{\mathbf{p},n}|^{\beta}.$$

Define the log-variations by

$$V_{n,0} = \frac{1}{(2^n - 1)^d} \sum_{\mathbf{p}=1}^{2^n-1} \log \left| X_{H,\alpha} \left(\frac{1+\mathbf{p}}{2^n} \right) - X_{H,\alpha} \left(\frac{\mathbf{p}}{2^n} \right) \right|.$$

Variations of processes are classical tools to perform identification of parameters: quadratic variations have been introduced for a while for Gaussian processes. The main result of this section concerns the asymptotic behavior of the log and β-variations and is given by the following theorem.

Theorem 5.2. *Let us assume that* $0 < H < 1$ *and* $0 < \widetilde{H} < 1$.

- *Convergence of* log-*variations.*

$$\lim_{n\to+\infty} -\frac{1}{n\log 2} V_{n,0} \stackrel{(a.s.)}{=} \widetilde{H}. \tag{104}$$

- *Convergence of* β-*variations.*

  – *For* $0 < \beta < \alpha$, *there exists a constant* $C_\beta > 0$ *such that:*

$$\lim_{n\to+\infty} 2^{n\beta\widetilde{H}} V_{n,\beta} \stackrel{(a.s.)}{=} C_\beta.$$

  – *For* $\alpha < \beta < 2$, *there exists a constant* $C_\beta > 0$ *such that:*

$$\lim_{n\to+\infty} 2^{n\beta\left(H+\frac{d}{\beta}-\frac{d}{2}\right)} V_{n,\beta} \stackrel{(a.s.)}{=} C_\beta.$$

The identification of fractional indexes H and $\widetilde{H}$ can then be performed as follows. A consistent estimator of $\widetilde{H}$ is given by

$$\widetilde{H}_n = -\frac{1}{\log 2^n} V_{n,0}. \tag{105}$$

To estimate H, we have to assume weak a priori knowledge on α, for instance that α belongs to the interval $]0, \alpha_{\text{sup}}[$, with $\alpha_{\text{sup}} < 2$ known. For any $\alpha_{\text{sup}} \leq \beta < 2$, a consistent estimator of H is then given by:

$$H_n = \frac{1}{\beta} \left(\log_2 \frac{V_{n-1,\beta}}{V_{n,\beta}} + \frac{\beta d}{2} - d \right). \tag{106}$$

Using the results (105) and (106), a consistent estimator of α is of course

$$\alpha_n = \frac{d}{\widetilde{H}_n - H_n + \frac{d}{2}}.$$

Please note that we could have estimated α using the results on the convergence of the β-variations. Actually if we assume that we know $(\beta, \log_2(\frac{V_{n-1,\beta}}{V_{n,\beta}}))$ for different values of β then α is the point on which the slope is changing. Although this method does not theoretically require any a priori knowledge for α, we believe it is not numerically feasible to determine a sampling design for the β's without this a priori knowledge.

Proof. To show the convergence of the log-variations, let us introduce notations. We start with $\widetilde{H}$-rescaled increments $Z_{\widetilde{H}}$ of $X_{H,\alpha}$

$$Z_{\widetilde{H}}(t, t') = \frac{X_{H,\alpha}(t') - X_{H,\alpha}(t)}{\|t' - t\|^{\widetilde{H}}} \qquad \forall t, t' \in \mathbb{R}^d. \tag{107}$$

We will show in the appendix the following lemma.

Lemma 5.3. *There exists a function* $\psi \in L^1(\mathbb{R}^d)$

$$\forall t, t' \in [0,1]^d, \ t \neq t', \ \forall \lambda \in \mathbb{R}, \quad \left| \mathbf{E}\left(e^{i\lambda Z_H(t,t')} \right) \right| \leq \psi(\lambda). \tag{108}$$

Moreover, for $0 < \beta < \alpha$,

$$\sup_{(t,t')\in[0,1]^d, t\neq t'} \mathbf{E}|Z_{\widetilde{H}}(t,t')|^\beta < \infty. \tag{109}$$

As a consequence we have the following result.

Lemma 5.4. *The family* $\log^2 |Z_H(t,t')|$, $t, t' \in [0,1]^d$, $t \neq t'$ *is uniformly integrable.*

Proof. In this proof we take $0 < \beta < \alpha$. Consider the following convex functions:

$$\phi_\beta(x) = e^\beta x \mathbf{1}_{x \leq 1} + e^{\beta \sqrt{x}} \mathbf{1}_{x > 1}.$$

Clearly, $\phi_\beta(x)/x \to +\infty$ when $x \to +\infty$. One has:

$$\mathbf{E}\phi_\beta(\log^2|Z_H(t,t')|) \leq e^\beta + \mathbf{E}|Z_H(t,t')|^\beta + \mathbf{E}|Z_H(t,t')|^{-\beta}.$$

Since for $0 < \beta < \alpha$, $\sup_{t,t'} \mathbf{E}|Z_H(t,t')|^\beta < +\infty$ it follows that the family of $\log^2|Z_H(t,t')|$ is uniformly integrable if $\sup_{t,t'} \mathbf{E}|Z_H(t,t')|^{-\beta}$ is finite for some $0 < \beta < \alpha$. Let us show this point. Since $\psi \in L^1(\mathbf{R}^d)$ because of (108), $Z_H(t,t')$ has a continuous density denoted by $p_{t,t'}$. Then, let $\mathcal{B}$ be the unit ball in $\mathbf{R}^d$

$$\begin{aligned}
\mathbf{E}|Z_H(t,t')|^{-\beta} &= \int_{\mathbf{R}^d} \|x\|^{-\beta} p_{t,t'}(x)dx \\
&= \int_B \|x\|^{-\beta} p_{t,t'}(x)dx + \int_{\mathbf{R}^d \setminus B} \|x\|^{-\beta} p_{t,t'}(x)dx \\
&\leq \|\psi\|_{L^1(\mathbf{R}^d)} \int_B \|x\|^{-\beta} dx + 1,
\end{aligned}$$

where we have bounded the Fourier transform of $p_{t,t'}$ by ψ. Hence

$$\sup_{t,t'} \mathbf{E}|Z_H(t,t')|^{-\beta} < +\infty$$

for every $0 < \beta < d$. $\qquad\qquad\qquad\qquad\qquad\qquad\qquad\qquad\qquad\qquad\square$

Let

$$W_{n,0} = \frac{1}{(2^n-1)^d} \sum_{\mathbf{p}=1}^{2^n-1} \log\left(2^{n\tilde{H}}\left|X_{H,\alpha}\left(\frac{1+\mathbf{p}}{2^n}\right) - X_{H,\alpha}\left(\frac{\mathbf{p}}{2^n}\right)\right|\right).$$

Then we will show

$$\lim_{n \to +\infty} \frac{W_{n,0}}{n} \stackrel{(a.s.)}{=} 0.$$

One has

$$\mathbf{E}W_{n,0} = \frac{1}{(2^n-1)^d} \sum_{\mathbf{p}=1}^{2^n-1} \mathbf{E}\log\left(2^{n\tilde{H}}\left|X_{H,\alpha}\left(\frac{1+\mathbf{p}}{2^n}\right) - X_{H,\alpha}\left(\frac{\mathbf{p}}{2^n}\right)\right|\right).$$

Since the family $\log^2|Z_H(t,t')|$, $t,t' \in [a,b]$, $t \neq t'$ is uniformly integrable, one has:

$$\sup_{\substack{n \geq 1, \\ \mathbf{p}=1,\ldots,2^n-1}} \mathbf{E}\log^2\left(2^{n\tilde{H}}\left|X_{H,\alpha}\left(\frac{1+\mathbf{p}}{2^n}\right) - X_{H,\alpha}\left(\frac{\mathbf{p}}{2^n}\right)\right|\right) < +\infty, \quad (110)$$

so that

$$\sup_{\substack{n \geq 1, \\ p=1,\ldots,2^n-1}} \mathbf{E} \left| \log \left(2^{n\tilde{H}} \left| X_{H,\alpha} \left(\frac{1+p}{2^n} \right) - X_{H,\alpha} \left(\frac{p}{2^n} \right) \right| \right) \right| < +\infty. \quad (111)$$

It follows

$$\lim_{n \to +\infty} \frac{\mathbf{E} W_{n,0}}{n} = 0.$$

If we denote by

$$A_\mathbf{p} = \log \left(2^{n\tilde{H}} \left| X_{H,\alpha} \left(\frac{1+p}{2^n} \right) - X_{H,\alpha} \left(\frac{p}{2^n} \right) \right| \right)$$

$$- \mathbf{E} \log \left(2^{n\tilde{H}} \left| X_{H,\alpha} \left(\frac{1+p}{2^n} \right) - X_{H,\alpha} \left(\frac{p}{2^n} \right) \right| \right),$$

the variance of $W_{n,0}$ is given by

$$\operatorname{var} W_{n,0} = \frac{1}{(2^n-1)^{2d}} \sum_{\mathbf{p,p'}=1}^{2^{n}-1} \mathbf{E} \left(A_\mathbf{p} A_{\mathbf{p'}} \right).$$

By Cauchy–Schwarz inequality,

$$\operatorname{var} W_{n,0}$$

$$\leq \frac{1}{(2^n-1)^{2d}} \left\{ \sum_{\mathbf{p}=1}^{2^{n}-1} \sqrt{\operatorname{var} \log \left(2^{n\tilde{H}} \left| X_{H,\alpha} \left(\frac{1+p}{2^n} \right) - X_{H,\alpha} \left(\frac{p}{2^n} \right) \right| \right)} \right\}^2.$$

Because of (110) and (111),

$$\sup_{n \geq 1} \operatorname{var} W_{n,0} < \infty. \quad (112)$$

Since

$$\mathbf{P} \left(\left| \frac{W_{n,0}}{n} - \frac{\mathbf{E} W_{n,0}}{n} \right| > a \right) \leq \frac{\sup_{n \geq 1} \operatorname{var} W_{n,0}}{n^2 a^2},$$

for every $a > 0$, Borel–Cantelli's lemma implies:

$$\lim_{n \to +\infty} \frac{W_{n,0}}{n} \overset{(a.s.)}{=} 0.$$

Definitions of $W_{n,0}$ and $V_{n,0}$ lead to

$$\frac{W_{n,0}}{n} = \tilde{H}\log 2 - V_{n,0},\tag{113}$$

and (104) is proved.

Integral representations of power functions are used extensively and are given in the following.

For all $0 < \beta < 2, \forall x \in \mathbb{R}$

$$|x|^\beta = \left(\int_\mathbb{R} \frac{e^{iy} - 1 - iy\mathbf{1}_{|y|\le 1}}{|y|^{1+\beta}}dy\right)^{-1}\int_\mathbb{R} \frac{e^{ixy} - 1 - ixy\mathbf{1}_{|y|\le 1}}{|y|^{1+\beta}}dy.$$

Because of the previous integral representation the following process:

$$S_n(y) = \frac{1}{(2^n - K)^d}\sum_{\mathbf{p}=1}^{2^n-\mathbf{K}}\exp\left(iy2^{n\tilde{H}}\Delta X_{\mathbf{p},n}\right), \ y \in \mathbb{R}$$

is introduced for the study of the β-variations and log-variations. Let

$$\Delta G_{\mathbf{p},n}(s) = \sum_{\ell=0}^K a_\ell G\left(\frac{\mathbf{p}+\ell}{2^n}, s\right),$$

where $G = \|t - s\|^{H-d/2} - \|s\|^{H-d/2}$ and

$$S(y) = \exp\left\{|y|^\alpha\int_{\mathbb{R}^d\times\mathbb{R}}[\exp(iv\Delta G_{0,1}(\sigma)) - 1 - iv\Delta G_{0,1}(\sigma)\mathbf{1}_{|v|\le 1}]d\sigma\frac{dv}{|v|^{1+\alpha}}\right\}.$$

We first prove the following intermediate lemma on $S_n(y)$.

Lemma 5.5.

$$\lim_{n\to+\infty}S_n(y) \overset{(a.s.)}{=} S(y).$$

The proof of Lemma 5.5 is postponed to the appendix.

We can now prove the convergence of the β-variations for $0 < \beta < \alpha$. The integral representation of power functions leads to:

$$\frac{2^{n\beta\tilde{H}}}{(2^n - K)^d}\sum_{\mathbf{p}=1}^{2^n-\mathbf{K}}|\Delta X_{\mathbf{p},n}|^\beta$$

$$= \int_\mathbb{R}\frac{S_n(y) - 1 - iy\mathbf{1}_{|y|\le 1}\frac{1}{(2^n-K)^d}\sum_{\mathbf{p}=1}^{2^n-\mathbf{K}}2^{n\tilde{H}}\Delta X_{\mathbf{p},n}}{|y|^{1+\beta}}dy.$$

The sequence $\sum_{p=1}^{2^n-K} \Delta X_{\mathbf{p},n}$ is a telescopic one: $\mathbb{E}\left(\sum_{p=1}^{2^n-K} \Delta X_{\mathbf{p},n}\right)^2$ converges to zero and can be overestimated by a constant. By Borel–Cantelli's lemma, $\dfrac{2^{n\tilde{H}}}{(2^n-K)^d}\sum_{p=1}^{2^n-K}\Delta X_{\mathbf{p},n}$ converges (a.s.) to 0.

An application of the dominated convergence theorem leads to:

$$\lim_{n\to+\infty}\frac{2^{n\beta\tilde{H}}}{(2^n-K)^d}\sum_{p=1}^{2^n-K}|\Delta X_{\mathbf{p},n}|^\beta \overset{(a.s.)}{=} \int_{\mathbb{R}}\frac{S(y)-1}{|y|^{1+\beta}}\,dy.$$

We now study the β-variations for $\alpha < \beta < 2$. The integral $\displaystyle\int_{\mathbb{R}}\frac{S(y)-1}{|y|^{1+\beta}}\,dy$ is divergent, of course the dominated convergence theorem cannot be applied anymore. First let us recall that

$$\int_{\mathbb{R}}\frac{\mathbb{E}S_n(y)-1}{|y|^{1+\beta}}\,dy = \int_{\mathbb{R}}\frac{\exp\left\{|y|^\alpha\int_{\mathbb{R}^d\times\mathbb{R}}E(v,\sigma)d\sigma\frac{dv}{|v|^{1+\alpha}}\mathbf{1}_{|v|\le|y|2^{n\frac{d}{\alpha}}}\right\}-1}{|y|^{1+\beta}}\,dy,$$

where

$$E(v,\sigma)=\exp(iv\Delta G_{0,1}(\sigma))-1-iv\Delta G_{0,1}(\sigma).$$

The previous integral is split into three terms. To make it short, the integrand with respect to y has not been written in the following when no confusion is possible.

- $|y|\le 2^{-n\frac{d}{\alpha}}$.

The change of variables $z=y2^{n\frac{d}{\alpha}}$ leads to

$$\int_{|y|\le 2^{-n\frac{d}{\alpha}}}\frac{\exp\left\{|y|^\alpha\int_{\mathbb{R}^d\times\mathbb{R}}E(v,\sigma)d\sigma\frac{dv}{|v|^{1+\alpha}}\mathbf{1}_{|v|\le|y|2^{n\frac{d}{\alpha}}}\right\}-1}{|y|^{1+\beta}}\,dy$$

$$=2^{nd\frac{\beta}{\alpha}}\int_{|z|\le 1}\frac{\exp\left\{2^{-nd}|z|^\alpha\int_{\mathbb{R}^d\times\mathbb{R}}E(v,\sigma)d\sigma\frac{dv}{|v|^{1+\alpha}}\mathbf{1}_{|v|\le|z|}\right\}-1}{|z|^{1+\beta}}\,dz.$$

Since $2^{-nd}\to 0$ as $n\to+\infty$, a Taylor expansion of order 1 is used:

$$\exp\left\{2^{-nd}|z|^\alpha\int_{\mathbb{R}^d\times\mathbb{R}}E(v,\sigma)d\sigma\frac{dv}{|v|^{1+\alpha}}\mathbf{1}_{|v|\le|z|}\right\}-1$$

$$=2^{-nd}|z|^\alpha\int_{\mathbb{R}^d\times\mathbb{R}}E(v,\sigma)d\sigma\frac{dv}{|v|^{1+\alpha}}\mathbf{1}_{|v|\le|z|}(1+o(1)).$$

Note that, because of the term $1_{|v|\leq|z|}$, the integral

$$\int_{|z|\leq 1}\frac{dz}{|z|^{1+\beta-\alpha}}\int_{\mathbb{R}^d\times\mathbb{R}}E(v,\sigma)d\sigma\frac{dv}{|v|^{1+\alpha}}1_{|v|\leq|z|}$$

is convergent. It follows that

$$\int_{|y|\leq 2^{-n\frac{d}{\alpha}}}\frac{\exp\left\{|y|^\alpha\int_{\mathbb{R}^d\times\mathbb{R}}E(v,\sigma)d\sigma\frac{dv}{|v|^{1+\alpha}}1_{|v|\leq|y|2^{n\frac{d}{\alpha}}}\right\}-1}{|y|^{1+\beta}}dy$$

$$=2^{n\left(-d+\frac{d\beta}{\alpha}\right)}\int_{|z|\leq 1}\frac{dz}{|z|^{1+\beta-\alpha}}\int_{\mathbb{R}^d\times\mathbb{R}}E(v,\sigma)d\sigma\frac{dv}{|v|^{1+\alpha}}1_{|v|\leq|z|}(1+o(1)).$$

- $|y|>\dfrac{1}{n}$.

  Because of the symmetry of $\dfrac{dv}{|v|^{1+\alpha}}$, the integral

  $$\int_{\mathbb{R}^d\times\mathbb{R}}E(v,\sigma)d\sigma\frac{dv}{|v|^{1+\alpha}}1_{|v|\leq|y|2^{n\frac{d}{\alpha}}}\text{ is negative.}$$

  We can bound $\exp\left\{|y|^\alpha\int_{\mathbb{R}^d\times\mathbb{R}}E(v,\sigma)d\sigma\frac{dv}{|v|^{1+\alpha}}1_{|v|\leq|y|2^{n\frac{d}{\alpha}}}\right\}$ by 1, so that

  $$\int_{|y|\geq 1/n}\frac{\exp\left\{|y|^\alpha\int_{\mathbb{R}^d\times\mathbb{R}}E(v,\sigma)d\sigma\frac{dv}{|v|^{1+\alpha}}1_{|v|\leq|y|2^{n\frac{d}{\alpha}}}\right\}-1}{|y|^{1+\beta}}dy\quad\leq Cn^\beta.$$

- $2^{-n\frac{d}{\alpha}}<|y|\leq\dfrac{1}{n}$.

  Since $\dfrac{1}{n}\to 0$, a Taylor expansion of order 1 leads to:

  $$\int_{2^{-n\frac{d}{\alpha}}\leq|y|\leq 1/n}\frac{\exp\left\{|y|^\alpha\int_{\mathbb{R}^d\times\mathbb{R}}E(v,\sigma)d\sigma\frac{dv}{|v|^{1+\alpha}}1_{|v|\leq|y|2^{n\frac{d}{\alpha}}}\right\}-1}{|y|^{1+\beta}}dy$$

  $$=\int_{2^{-n\frac{d}{\alpha}}\leq|y|\leq 1/n}\frac{dy}{|y|^{1+\beta-\alpha}}\int_{\mathbb{R}^d\times\mathbb{R}}E(v,\sigma)d\sigma\frac{dv}{|v|^{1+\alpha}}1_{|v|\leq|y|2^{n\frac{d}{\alpha}}}(1+o(1)).$$

The change of variable $z=y2^{n\frac{d}{\alpha}}$ leads to:

$$\int_{2^{-n\frac{d}{\alpha}}\leq|y|\leq 1/n}\frac{dy}{|y|^{1+\beta-\alpha}}\int_{\mathbb{R}^d\times\mathbb{R}}E(v,\sigma)d\sigma\frac{dv}{|v|^{1+\alpha}}1_{|v|\leq|y|2^{n\frac{d}{\alpha}}}$$

$$=2^{n\left(-d+\frac{d\beta}{\alpha}\right)}\int_{1\leq|z|\leq 2^{n\frac{d}{\alpha}}/n}\frac{dz}{|z|^{1+\beta-\alpha}}\int_{\mathbb{R}^d\times\mathbb{R}}E(v,\sigma)d\sigma\frac{dv}{|v|^{1+\alpha}}1_{|v|\leq|z|}.$$

Since $\beta > \alpha$, the integral

$$\int_{1 \leq |z| \leq 2^{n\frac{d}{\alpha}}/n} \frac{dz}{|z|^{1+\beta-\alpha}} \int_{\mathbb{R}^d \times \mathbb{R}} E(v,\sigma)d\sigma \frac{dv}{|v|^{1+\alpha}} \mathbf{1}_{|v| \leq |z|}$$

converges to

$$\int_{1 \leq |z|} \frac{dz}{|z|^{1+\beta-\alpha}} \int_{\mathbb{R}^d \times \mathbb{R}} E(v,\sigma)d\sigma \frac{dv}{|v|^{1+\alpha}} \mathbf{1}_{|v| \leq |z|},$$

so that

$$\int_{2^{-n\frac{d}{\alpha}} \leq |y| \leq 1/n} \frac{\exp\left\{ |y|^\alpha \int_{\mathbb{R}^d \times \mathbb{R}} E(v,\sigma)d\sigma \frac{dv}{|v|^{1+\alpha}} \mathbf{1}_{|v| \leq |y| 2^{n\frac{d}{\alpha}}} \right\} - 1}{|y|^{1+\beta}} dy$$

$$= 2^{n\left(-d+\frac{d\beta}{\alpha}\right)}(1+o(1)) \int_{|z| \geq 1} \frac{dz}{|z|^{1+\beta-\alpha}} \int_{\mathbb{R}^d \times \mathbb{R}} E(v,\sigma)d\sigma \frac{dv}{|v|^{1+\alpha}} \mathbf{1}_{|v| \leq |z|}.$$

To sum up, the first term is equivalent to $C2^{n\left(-d+\frac{d\beta}{\alpha}\right)}$, the third is equivalent to $C2^{n\left(-d+\frac{d\beta}{\alpha}\right)}$ and the second one is negligible as compared to the two others. We have proved that

$$2^{n\left(d-\frac{d\beta}{\alpha}\right)} \int_{\mathbb{R}} \frac{\mathbb{E}S_n(y) - 1}{|y|^{1+\beta}} dy \to C.$$

From Lemma 5.5, $S_n(y) = \mathbb{E}S_n(y)(1 + o_{(a.s.)}(1))$. We have therefore proved that $2^{n\beta\widetilde{H}} 2^{n(d-d\beta/\alpha)} V_{n,\beta}$ converges, as $n \to +\infty$ to a constant. Since $\beta\widetilde{H} + d - d\beta/\alpha) = \beta(H - d/2 + d/\beta)$, Theorem 5.2 is proved. $\qquad \square$

6 Simulation

So far in this article we have seen many fractional fields, and this section will be devoted to the simulation of some of these fields. In the literature, there exist articles for simulation of the fractional fields that are non Gaussian. In [13] a wavelet expansion is used to approximate harmonizable and well-balanced type of fractional stable processes. For the linear fractional stable processes, which are fields in dimension 1 similar to mafsfs, the fast Fourier transform is the main tool for simulation in [37,42]. One can also quote [26], where another integral representation of the linear fractional stable processes is used to obtain simulation of the sample paths. Even though all these processes are stable, they have different distributions and for each one a specific method is used. Concerning non stable fields, generalized shot noise series introduced for simulations of Lévy processes in [31–33] were used for simulation of the sample paths of real harmonizable multifractional fields in [22]. One of the advantages of this method is the fact that it can be applied to fractional fields that are neither with stationary increments nor self-similar. Moreover, it is straightforward to apply this technique to the simulation of fields indexed by multidimensional spaces. In this section, our main goal is to show how this method can be applied to most of the fractional fields.

Let us describe how one can obtain an algorithm of simulation, when an integral representation of the fractional field is known. We will be interested in the simulation of stochastic integrals of the form

$$X^f(t) = \int_{\mathbb{R}^d} f(t, s)\, \Lambda(ds), \quad t \in \mathbb{R}^d,$$

with Λ a random measure, which is either a Lévy or a stable random measure defined respectively in Sects. 2.2 and 2.3.

Basically, one can consider the random measure Λ as a sum of weighted Dirac masses at random points at the arrival times of a standard Poisson process. After the transformation, the integrals are series which may be simulated by properly truncating the number of terms.

We also would like to stress that we have obtained rates of convergence for the truncating series. More precisely, almost sure rates of convergence are given both for each marginal of the field, and uniformly if the field is simulated on a compact set. The almost sure convergence is related to asymptotic developments of the deterministic kernel in the integral representation of the field. Let us also emphasize Theorem 6.1, which is an important tool to reach rates of convergence for series of symmetric random variables under moment assumptions. This theorem may have interest of its own and is needed in the heavy tail cases. Rates of convergence in L^r-norm with explicit constant are further obtained.

When the control measure of Λ has infinite mass, a technical complication arises. Following [2, 22], one part of X^f will then be approximated by a Gaussian field and the error due to this approximation will be given in terms of Berry–Esseen bounds. The other part will be represented as series.

In Sect. 6.1, rates of almost sure convergence for shot noise series are studied. Section 6.2 is devoted to some facts concerning stochastic integrals with respect to random measures, which are refinement of Sect. 2. Then, convergence and rates of convergence of the generalized shot noise series are given in Sect. 6.3. Section 6.4 gives an approximation of the stochastic integrals, when the control measure has infinite mass, and establishes Berry–Esseen bounds. Examples, that include most of the classical fractional fields, are given in Sect. 6.6, illustrated by simulations. Section 6.7 is devoted to the case of complex random measures, which are important for harmonizable fields. The proofs of Theorems 6.1 and 6.2 are given in the Sect. 6.7.

6.1 Rate of Almost Sure Convergence for Shot Noise Series

In this section, we first establish the main tools to reach rates of convergence of the approximation proposed in Sect. 6.3. The two following theorems study rates of convergence for series of symmetric random variables. In particular, they can be applied to

$$S_N^\gamma = \sum_{n=1}^{N} T_n^{-1/\gamma} X_n, \tag{114}$$

where $0 < \gamma < 2$ and T_n is the nth arrival time of a Poisson process with intensity 1. Let us recall that if $(X_n)_{n\geq1}$ is independent of $(T_n)_{n\geq1}$, the shot noise series (114) converge almost surely to a stable random variable with index γ as soon as (X_n), $n \geq 1$, are independent and identically distributed (i.i.d) L^γ-symmetric random variables, see for instance [23,34]. Under a stronger integrability assumption, a rate of almost sure convergence is given by Theorem 6.1. Theorem 6.2 gives a rate of absolute almost sure convergence.

Theorem 6.1. *Let $(X_n)_{n\geq1}$ be a sequence of i.i.d. symmetric random variables. Assume that $(X_n)_{n\geq1}$ is independent of $(T_n)_{n\geq1}$ and of a sequence $(Y_n)_{n\geq1}$ which satisfies*

$$|Y_n| \leq C T_n^{-1/\gamma} \quad a.s. \tag{115}$$

for some finite constants $C > 0$ and $\gamma \in (0, 2)$. Furthermore, assume $\mathbb{E}(|X_n|^r) < +\infty$ for some $r > \gamma$. Then, for every $\varepsilon \in (0, 1/\gamma - 1/(r \wedge 2))$, almost surely,

$$\sup_{N\geq1} N^\varepsilon \left| \sum_{n=N+1}^{+\infty} Y_n X_n \right| < +\infty.$$

Proof. See the Sect. 6.7.   □

The Theorem 6.1 will give us a rate of almost sure convergence of our approximation by generalized shot noise series (see Sect. 6.3). In this paper, we are also interested in the uniform convergence of our approximation, when the field X^f is simulated on a compact set. The next theorem will be the main tool to establish this uniform convergence and obtain a rate of uniform convergence.

Theorem 6.2. *Let $(X_n)_{n\geq1}$ be a sequence of i.i.d random variables and $\gamma \in (0, 1)$. Assume that $(X_n)_{n\geq1}$ is independent of $(T_n)_{n\geq1}$ and that $\mathbb{E}(|X_n|^r) < +\infty$ for some $r > \gamma$. Then, for every $\varepsilon \in (0, 1/\gamma - 1/(r \wedge 1))$, almost surely,*

$$\sup_{N\geq1} N^\varepsilon \sum_{n=N+1}^{+\infty} T_n^{-1/\gamma} |X_n| < +\infty.$$

Proof. See the Sect. 6.7.   □

6.2   Stochastic Integrals Revisited

In this section, the definitions in the Sect. 2.1 of stochastic integrals with respect to
Poisson random measures are generalized (see [30] for more details). Let $N(ds, dv)$
be a Poisson random measure on $\mathbb{R}^d \times \mathbb{R}$ with intensity $n(ds, dv) = ds\mu(dv)$.
Assume that the non-vanishing σ-finite measure $\mu(dv)$ is a symmetric measure such
that

$$\int_{\mathbb{R}} \left(|v|^2 \wedge 1 \right) \mu(dv) < +\infty, \tag{116}$$

where $a \wedge b = \min(a, b)$. In particular, $\mu(dv)$ may have an infinite second order
moment. The assumption (116) is weaker than the assumption (4). Similarly, in
Sect. 6.7, the control measure satisfies a weaker assumption than the one made
in (16).

The stochastic integral

$$\int_{\mathbb{R}^d \times \mathbb{R}} \varphi(s, v) \left[N(ds, dv) - (1 \vee |\varphi(s, v)|)^{-1} n(ds, dv) \right],$$

where $a \vee b = \max(a, b)$, is defined if and only if $\displaystyle\int_{\mathbb{R}^d \times \mathbb{R}} \left(|\varphi(s, v)|^2 \wedge 1 \right) n(ds, dv) <$
$+\infty$, see for instance Lemma 12.13 page 236 in [18].

Then, we can consider a random measure $\Lambda(ds)$ on $\mathbb{R}^d$ defined by

$$\int_{\mathbb{R}^d} g(s) \Lambda(ds) = \int_{\mathbb{R}^d \times \mathbb{R}} g(s)v \left(N(ds, dv) - (|g(s)v| \vee 1)^{-1} n(ds, dv) \right) \tag{117}$$

for every $g : \mathbb{R}^d \to \mathbb{R}$ such that $\displaystyle\int_{\mathbb{R}^d \times \mathbb{R}} |g(s)v|^2 \wedge 1 \, n(ds, dv) < +\infty$. We have that

$$\mathbb{E}\left[\exp\left(i \int_{\mathbb{R}^d} g(s) \Lambda(ds) \right) \right]$$

$$= \exp\left[\int_{\mathbb{R}^d \times \mathbb{R}} \left[\exp(ig(s)v) - 1 - ig(s)v \mathbf{1}_{|g(s)v| \leq 1} \right] ds \, \mu(dv) \right], \tag{118}$$

see for instance [18]. Therefore Λ is an infinitely divisible random measure.

As explained below (see Examples 6.3 and 6.4), Lévy random measures and
stable random measures are examples of such infinitely divisible random measures represented by a Poisson random measure owing to (117). Here are some
illustrations.

Example 6.3. Let $\mu(dv)$ be a symmetric measure such that

$$\int_{\mathbb{R}} |v|^2 \mu(dv) < +\infty.$$

Then, for every $g \in L^2(\mathbb{R}^d)$,

$$\int_{\mathbb{R}^d \times \mathbb{R}} |g(s)v| \mathbf{1}_{|g(s)v| \geq 1} \left(1 - \frac{1}{|g(s)v| \vee 1} \right) ds\mu(dv)$$

$$\leq \int_{\mathbb{R}} |v|^2 \mu(dv) \int_{\mathbb{R}^d} g(v)^2 dv < \infty.$$

Since $\mu(dv)$ is a symmetric measure

$$\int_{\mathbb{R}^d \times \mathbb{R}} g(s)v \left(1 - \frac{1}{|g(s)v| \vee 1} \right) ds\mu(dv) = 0$$

and (117) can be rewritten as

$$\int_{\mathbb{R}^d} g(s) \Lambda(ds) = \int_{\mathbb{R}^d \times \mathbb{R}} g(s)v \left(N(ds, dv) - n(ds, dv) \right).$$

If the symmetric measure $\mu(dv)$ satisfies the assumptions (4), i.e. if

$$\forall p \geq 2, \ \int_{\mathbb{R}} |v|^p \mu(dv) < +\infty,$$

$\Lambda(ds)$ is a Lévy random measure represented by the Poisson random measure $N(ds, dv)$ in the sense of the Sect. 2.1. Under the above assumptions, the field $(X_H(t))_{t \in \mathbb{R}^d}$, defined by

$$X_H(t) = \int_{\mathbb{R}^d} \left(\|t - s\|^{H-d/2} - \|s\|^{H-d/2} \right) \Lambda(ds)$$

is a moving average fractional Lévy field, in short mafLf, with index H ($0 < H < 1$, $H \neq d/2$).

Example 6.4. In the case where

$$\mu(dv) = \frac{dv}{|v|^{1+\alpha}}$$

with $0 < \alpha < 2$, the random measure $\Lambda(ds)$, defined by (117), is a symmetric α-stable random measure in the sense of Sect. 2.3. Then, for instance,

$$X_H(t) = \int_{\mathbb{R}^d} \left(\|t - s\|^{H-d/\alpha} - \|s\|^{H-d/\alpha} \right) \Lambda(ds), \quad t \in \mathbb{R}^d$$

is a moving average fractional stable field, in short mafsf, with index H ($0 < H < 1$, $H \neq d/\alpha$).

In the following, we will be interested in the simulation of stochastic integrals of the form

$$X^f(t) = \int_{\mathbb{R}^d} f(t, s) \Lambda(ds), \quad t \in \mathbb{R}^d, \tag{119}$$

where $\Lambda(ds)$ is defined by (117) and $f : \mathbb{R}^d \times \mathbb{R}^d \to \mathbb{R}$ is such that for every $t \in \mathbb{R}^d$,

$$\int_{\mathbb{R}^d \times \mathbb{R}} \left(|f(t, s)v|^2 \wedge 1 \right) n(ds, dv) < +\infty. \tag{120}$$

To analyze these stochastic integrals, we represent them as series (known as shot noise series) for which we carefully study the rates of convergence.

6.3 Generalized Shot Noise Series

An overview of representations of infinitely divisible laws as series is given in [32, 33], and the field X^f is an infinitely divisible field. Such a representation of rhmLfs introduced in [21] has been studied in [22]. As in the case of rhmLfs, the infinitely divisible field X^f can be represented as a generalized shot noise series, as soon as the control measure $\mu(dv)$ has finite mass. Hence, in this section, we assume

$$\mu(\mathbb{R}) < +\infty. \tag{121}$$

Let us recall that $\mu(dv)$ is a non-vanishing measure, i.e. $\mu(\mathbb{R}) \neq 0$.

Let us now introduce some notation that will be used throughout the paper.

Notation. Let $(V_n)_{n \geq 1}$ and $(U_n)_{n \geq 1}$ be independent sequences of random variables. We assume that $(U_n, V_n)_{n \geq 1}$ is independent of $(T_n)_{n \geq 1}$.

- $(V_n)_{n \geq 1}$ is a sequence of i.i.d. random variables with common law $\mu(dv)/\mu(\mathbb{R})$.
- $(U_n)_{n \geq 1}$ is a sequence of i.i.d. random variables such that U_1 is uniformly distributed on the unit sphere S^{d-1} of the Euclidean space $\mathbb{R}^d$.
- c_d is the volume of the unit ball of $\mathbb{R}^d$.

The following statement is the main series representation, we will be using in our investigation.

Proposition 6.5. *Assume that* (120) *is fulfilled. Then, for every* $t \in \mathbb{R}^d$, *the series*

$$Y^f(t) = \sum_{n=1}^{+\infty} f\left(t, \left(\frac{T_n}{c_d \mu(\mathbb{R})} \right)^{1/d} U_n \right) V_n \tag{122}$$

converges almost surely. Furthermore, $\{X^f(t) : t \in \mathbb{R}^d\} \overset{(d)}{=} \{Y^f(t) : t \in \mathbb{R}^d\}$.

Remark 6.6. In the framework of rhmLfs, [22] directly establishes the almost convergence of the shot noise series in the space of continuous functions endowed with the topology of uniform convergence on compact sets. Such a result assumes the continuity of the deterministic kernel f and in our framework, this kernel function may be discontinuous. Nevertheless, under assumptions on the asymptotic expansion of f as $\|s\|$ tends to infinity, (122) also converges almost surely, uniformly in t, on each compact set. Such a result, stated in Proposition 6.10, will be deduced from the Theorem 6.2.

Proof. Let p be an integer, $p \geq 1$, $(u_1, \cdots, u_p) \in \mathbb{R}^p$ and $(t_1, \cdots, t_p) \in (\mathbb{R}^d)^p$. We consider the Borel measurable map

$$H :]0, +\infty[\times \mathscr{D} \longrightarrow \mathbb{R}$$

$$(r, \tilde{v}) \longmapsto \sum_{j=1}^{p} u_j f\left(t_j, \left(\frac{r}{c_d \mu(\mathbb{R})}\right)^{1/d} u\right) v,$$

where $\mathscr{D} = S^{d-1} \times \mathbb{R}$ and $\tilde{v} = (u, v) \in \mathscr{D}$ means $u \in S^{d-1}$ and $v \in \mathbb{R}$. Then, define a measure Q on the Borel σ-field $\mathscr{B}(\mathbb{R})$ by

$$\forall B \in \mathscr{B}(\mathbb{R}), \; Q(B) = \int_0^{+\infty} \int_{\mathscr{D}} \mathbf{1}_{B \setminus \{0\}}(H(r, \tilde{v})) \, \lambda(d\tilde{v}) \, dr,$$

where λ is the law of $\widetilde{V_n} = (U_n, V_n)$. In the previous formula, please note that the measure $Q(\{0\}) = 0$, which is a necessary condition for a Lévy measure. Then,

$$\int_{\mathbb{R}} |y|^2 \wedge 1 \, Q(dy)$$

$$= \int_{]0, +\infty[\times \mathscr{D}} H^2(r, \tilde{v}) \wedge 1 \, dr \, \lambda(d\tilde{v})$$

$$= \int_0^\infty dr \int_{S^{d-1}} du \int_{-\infty}^\infty \frac{\mu(dv)}{\mu(\mathbb{R})} [(\sum_{j=1}^{p} u_j f\left(t_j, \left(\frac{r}{c_d \mu(\mathbb{R})}\right)^{1/d} u\right) v)^2 \wedge 1]$$

Then, proceeding as in the proof of Proposition 3.1 in [22], i.e. using the change of variable $\rho = (r/(c_d \mu(\mathbb{R})))^{1/d}$ and polar coordinates, one obtains that the previous integral is equal to

$$d c_d \mu(\mathbb{R}) \int_{\mathbb{R}} d\rho \rho^{d-1} \int_{S^{d-1}} du \int_{-\infty}^\infty \frac{\mu(dv)}{\mu(\mathbb{R})} [(\sum_{j=1}^{p} u_j f(t_j, \rho)v)^2 \wedge 1]$$

$$= d c_d \int_{\mathbb{R}^d} ds \int_{-\infty}^\infty \mu(dv)[((\sum_{j=1}^{p} u_j f(t_j, \rho u))^2 v^2) \wedge 1] \, .$$

Hence it is bounded by $K \sum_{j=1}^{p} \int_{\mathbb{R}^d \times \mathbb{R}} [(u_j f(t_j, s)^2) v^2 \wedge 1] ds \mu(dv) < +\infty$, where $0 < K < +\infty$.

Since $\int_{\mathbb{R}} |y|^2 \wedge 1 \, Q(dy) < +\infty$, Q is a Lévy measure on $\mathbb{R}$. Therefore, according to Theorem 2.4 in [32], the sequence

$$\sum_{n=1}^{N} H\left(T_n, \widetilde{V_n}\right) - A(T_N),$$

where for $s \geq 0$,

$$A(s) = \int_0^s \int_{\mathscr{D}} H(r, \widetilde{v}) \mathbf{1}_{|H(r,\widetilde{v})| \leq 1} \, \lambda(d\widetilde{v}) \, dr,$$

converges almost surely as $N \to +\infty$. Moreover, since μ is a finite and symmetric measure, by the definition of H and of the measure $\lambda(d\widetilde{v})$, $A(s) = 0$ for every $s \geq 0$. Therefore, (taking $p = 1$), for every t,

$$Y^f(t) = \sum_{n=1}^{+\infty} f\left(t, \left(\frac{T_n}{c_d \mu(\mathbb{R})}\right)^{1/d} U_n\right) V_n$$

converges almost surely. Furthermore, due to Theorem 2.4 in [32], we have that

$$\mathbb{E}\left[\exp\left(i \sum_{j=1}^{p} u_j Y^f(t_j)\right)\right] = \exp\left[\int_{\mathbb{R}} (\exp(iy) - 1 - iy\mathbf{1}_{|y| \leq 1}) \, Q(dy)\right]$$

By the definition of Q and the symmetry of $\mu(dv)$, one easily sees that $\{X^f(t) : t \in \mathbb{R}^d\} \stackrel{(d)}{=} \{Y^f(t) : t \in \mathbb{R}^d\}$. The proof of Proposition 6.5 is then complete. $\quad\square$

On the basis of Proposition 6.5, Y^f, which is equal in law to X^f, is approximated by

$$Y_N^f(t) = \sum_{n=1}^{N} f\left(t, \left(\frac{T_n}{c_d \mu(\mathbb{R})}\right)^{1/d} U_n\right) V_n, \quad t \in \mathbb{R}^d. \tag{123}$$

We now explain in a few words how the rate of convergence of Y_N^f to Y^f can be studied. Firstly, let us recall the following classical result for Poisson arrival times:

$$\lim_{n \to +\infty} \frac{T_n}{n} = 1 \text{ almost surely.} \tag{124}$$

Hence, the asymptotics of (123) depends on $(V_n)_{n \geq 1}$ and on the asymptotics of $f(t, s)$ as $\|s\|$ tends to infinity. Under an assumption on this asymptotics, the rate of convergence of Y_N^f will be deduced from the rate of convergence of some series of the kind of S_N^γ defined by (114).

Let us first study the almost sure and L^r errors for each fixed t.

Theorem 6.7. *Let $t \in \mathbb{R}^d$. Assume that*

$$\forall s \neq 0, \ |f(t,s)| \leq \frac{C}{\|s\|^{\beta}}, \tag{125}$$

where $\beta > d/2$ and $C > 0$. Furthermore, assume there exists $r \in (d/\beta, 2]$ such that $\mathbb{E}(|V_1|^r) < +\infty$

1. Then, for every $\varepsilon \in (0, \beta/d - 1/r)$, almost surely,

$$\sup_{N \geq 1} N^{\varepsilon} \left| Y^f(t) - Y_N^f(t) \right| < +\infty.$$

2. Moreover, for every integer $N > r\beta/d$,

$$\mathbb{E}\left(\left| Y_N^f(t) - Y^f(t) \right|^r \right) \leq C(r,\beta) \frac{D(N,r,\beta)}{N^{r\beta/d-1}}, \tag{126}$$

where

$$D(N,r,\beta) = \frac{\Gamma(N+1-r\beta/d)\,(N+1)^{r\beta/d}}{\Gamma(N+1)} \tag{127}$$

and

$$C(r,\beta) = \frac{dC^r\,(c_d\mu(\mathbb{R}))^{r\beta/d}\mathbb{E}(|V_1|^r)}{r\beta - d}. \tag{128}$$

Remark 6.8. Remark that $\lim_{N \to +\infty} D(N,r,\beta) = 1$ by the Stirling formula. Hence, Theorem 6.7 gives a rate of convergence in L^r for the series Y_N^f. Furthermore, (126) allows us to control the error of approximation in simulation.

Remark 6.9. Assume that (125) is only fulfilled for $\|s\| \geq A$. Then, let

$$g(t,s) = f(t,s)\mathbf{1}_{\|s\| \geq A}$$

and remark that

$$Y^f = Y^g + Y^{f-g}, \tag{129}$$

where Y^h is associated with h by (122). Hence, since g satisfies the assumptions of Proposition 6.7, an almost sure or L^r error may be obtained. Furthermore, in view of (124),

$$Y^{f-g}(t) = \sum_{n=1}^{+\infty} (f - g)\left(t, \left(\frac{T_n}{c_d\mu(\mathbb{R})}\right)^{1/d} U_n\right) V_n$$

is, almost surely, a finite sum since for n large enough, $T_n > A^d c_d\mu(\mathbb{R})$. This remark is used for mafsfs or mafLfs in Sect. 6.6.

Let us now prove Theorem 6.7.

Proof. In the following,

$$s_n = \left(\frac{T_n}{c_d \mu(\mathbb{R})} \right)^{1/d} U_n.$$

1. Proof of Part 1: Rate of Almost Sure Convergence
In view of (125),

$$|f(t, s_n)| \leq \frac{C(c_d \mu(\mathbb{R}))^{\beta/d}}{T_n^{\beta/d}}. \tag{130}$$

Then, by applying Theorem 6.1 with $T_n = V_n$ and $Y_n = f(t, s_n)$,

$$\sup_{N \geq 1} N^\varepsilon \left| Y^f(t) - Y_N^f(t) \right| < +\infty \text{ almost surely,}$$

for every $\varepsilon \in (0, \beta/d - 1/r)$.

2. Proof of Part 2: L^r-error
Let $\varepsilon_n, n \geq 1$ be a sequence of independent Bernoulli random variables taking values ± 1 with probability $1/2$. Then, for every $r \in (0, 2]$ and every real numbers $a_1, \ldots, a_n$,

$$\mathbb{E}\left(\left| \sum_{i=1}^n a_i \varepsilon_i \right|^r \right) \leq \sum_{i=1}^n |a_i|^r.$$

Indeed, by Jensen's inequality

$$\mathbb{E}\left(\left| \sum_{i=1}^n a_i \varepsilon_i \right|^r \right) \leq \left[\mathbb{E}\left(\left| \sum_{i=1}^n a_i \varepsilon_i \right|^2 \right) \right]^{r/2} = \left(\sum_{i=1}^n |a_i|^2 \right)^{r/2}$$

and the result follows since $(a + b)^{r/2} \leq a^{r/2} + b^{r/2}$ ($r \in (0, 2]$) for every $a, b \geq 0$. Now, $V_n, n \geq 1$, is a sequence of independent symmetric random variables. Thus, it has the same distribution as $\varepsilon_n V_n, n \geq 1$ where $\varepsilon_n, n \geq 1$ is assumed to be independent of $V_n, n \geq 1$, as well as of the sequence $s_n, n \geq 1$. Therefore, conditionally on V_n and s_n, it follows from the latter that

$$\mathbb{E}\left(\left| \sum_{n=N+1}^P f(t, s_n) V_n \right|^r \right) \leq \sum_{n=N+1}^P \mathbb{E}(|f(t, s_n)|^r |V_n|^r)$$

$$= \mathbb{E}(|V_1|^r) \sum_{n=N+1}^P \mathbb{E}(|f(t, s_n)|^r).$$

Then, by (130),

$$\mathbb{E}\left(\left|\sum_{n=N+1}^{P} f(t,s_n)V_n\right|^r\right) \leq C^r (c_d\mu(\mathbb{R}))^{r\beta/d}\mathbb{E}(|V_1|^r)\sum_{n=N+1}^{P}\mathbb{E}\left(T_n^{-r\beta/d}\right)$$

$$\leq C^r (c_d\mu(\mathbb{R}))^{r\beta/d}\mathbb{E}(|V_1|^r)\sum_{n=N+1}^{P}\frac{\Gamma(n-r\beta/d)}{\Gamma(n)}.$$

Therefore,

$$\mathbb{E}\left(\left|\sum_{n=N+1}^{P} f(t,s_n)V_n\right|^r\right)$$

$$\leq C^r (c_d\mu(\mathbb{R}))^{r\beta/d}\mathbb{E}(|V_1|^r)\sup_{n\geq N} D(n,r,\beta)\sum_{n=N+1}^{+\infty}\frac{1}{n^{r\beta/d}}$$

where $D(n,r,\beta)$ is defined by (127). According to the proof of Proposition 3.2 in [22],

$$\sup_{n\geq N} D(n,r,\beta) = D(N,r,\beta)$$

and then

$$\mathbb{E}\left(\left|\sum_{n=N+1}^{P} f(t,s_n)V_n\right|^r\right) \leq \frac{dC^r (c_d\mu(\mathbb{R}))^{r\beta/d}\mathbb{E}(|V_1|^r)D(N,r,\beta)}{(r\beta-d)N^{r\beta/d-1}}$$

since $r > d/\beta$. Then, by the Fatou lemma,

$$\mathbb{E}\left(\left|\sum_{n=N+1}^{+\infty} f(t,s_n)V_n\right|^r\right) \leq \frac{dC^r (c_d\mu(\mathbb{R}))^{r\beta/d}\mathbb{E}(|V_1|^r)D(N,r,\beta)}{(r\beta-d)N^{r\beta/d-1}}.$$

The proof of Theorem 6.7 is complete. □

Actually, if f admits an expansion, roughly speaking uniform in t, as $\|s\|$ tends to infinity, the next theorem gives a rate of uniform convergence in t for Y_N^f.

Theorem 6.10. *Let $K \subset \mathbb{R}^d$ be a compact set, $p \geq 1$ and $(\beta_i)_{1\leq i\leq p}$ be a non-decreasing sequence. such that $\beta_1 > d/2$ and $\beta_p > d$. Assume that for every $t \in K$ and $s \neq 0$,*

$$\left|f(t,s) - \sum_{j=1}^{p-1}\frac{a_j(t)b_j(s/\|s\|)}{\|s\|^{\beta_j}}\right| \leq \frac{b_p(s/\|s\|)}{\|s\|^{\beta_p}} \tag{131}$$

where a_j, $j = 1, \ldots, p-1$, *are real-valued continuous functions. Furthermore, as-*
sume that there exists $r \in (d/\beta_1, 2]$ *such that* $\mathbb{E}(|V_n|^r) < +\infty$ *and* $\mathbb{E}(|b_j(U_n)|^r) <$
$+\infty$ *for* $j = 1, \ldots, p$. *Then for every* $\varepsilon \in (0, \min(\beta_1/d - 1/r, \beta_p/d - 1/(1 \wedge r)))$,

$$\sup_{N \geq 1} N^\varepsilon \sup_{t \in K} \left| Y^f(t) - Y_N^f(t) \right| < +\infty \text{ almost surely.}$$

Remark 6.11. In (131), the functions b_j provide an anisotropic control of the asymptotic expansion of f.

Proof. We have

$$\left| Y^f(t) - Y_N^f(t) \right| \leq \sum_{j=1}^{p-1} |a_j(t)| \left| \sum_{n=N+1}^{+\infty} \left(\frac{T_n}{c_d \mu(\mathbb{R})} \right)^{-\beta_j/d} b_j(U_n) V_n \right|$$

$$+ \sum_{n=N+1}^{+\infty} \left(\frac{T_n}{c_d \mu(\mathbb{R})} \right)^{-\beta_p/d} |b_p(U_n) V_n|.$$

Note that $(b_j(U_n)V_n)_{n \geq 1}$ are i.i.d. symmetric random variables such that $\mathbb{E}(|b_j(U_n)V_n|^r) < +\infty$. Hence, since $0 < d/\beta_j < r \leq 2$, by Theorem 6.1, for every $\varepsilon \in (0, \beta_j/d - 1/r)$,

$$\sup_{N \geq 1} N^\varepsilon \left| \sum_{n=N+1}^{+\infty} T_n^{-\beta_j/d} b_j(U_n) V_n \right| < +\infty \text{ almost surely.}$$

In addition, since $\mathbb{E}(|b_p(U_n)V_n|^r) < +\infty$ and $d/\beta_p < 1$, by Theorem 6.2, for every $\varepsilon \in (0, \beta_p/d - 1/(1 \wedge r))$,

$$\sup_{N \geq 1} N^\varepsilon \sum_{n=N+1}^{+\infty} T_n^{-\beta_p/d} |b_p(U_n) V_n| < +\infty \text{ almost surely,}$$

which ends the proof since a_j, $j = 1, \ldots, p-1$, are continuous and thus bounded on the compact set K. □

6.4 Normal Approximation

When the assumption (121) is not fulfilled, the results of Sect. 6.3 cannot be directly applied. In this case, the simulation of X^f is not only based on a series expansion but also on a normal approximation. Actually, following [2, 22], we will split the field X^f into two fields $X_{\varepsilon,1}^f$ and $X_{\varepsilon,2}^f$. It leads to a decomposition of Λ into two random measures $\Lambda_{\varepsilon,1}$ and $\Lambda_{\varepsilon,2}$ such that the control measure of $\Lambda_{\varepsilon,2}$ satisfies the assumption (121). As a consequence of Sect. 6.3, $X_{\varepsilon,2}^f$ can be represented as a series. This section is thus devoted to the simulation of the first part $X_{\varepsilon,1}^f$ that will be handled by normal approximation of the Berry–Esseen type.

Suppose now that

$$\mu(\mathbb{R}) = +\infty, \tag{132}$$

which is the case for mafsfs. Then let $\varepsilon > 0$ and let us split

$$X^f = X_{\varepsilon,1}^f + X_{\varepsilon,2}^f$$

into two random fields where

$$X_{\varepsilon,1}^f(t) = \int_{\mathbb{R}^d \times \mathbb{R}} f(t,s)v\mathbf{1}_{|v|<\varepsilon} \left(N(ds,dv) - (|f(t,s)v| \vee 1)^{-1} n(ds,dv) \right) \tag{133}$$

and

$$X_{\varepsilon,2}^f(t) = \int_{\mathbb{R}^d \times \mathbb{R}} f(t,s)v\mathbf{1}_{|v|\geq\varepsilon} \left(N(ds,dv) - (|f(t,s)v| \vee 1)^{-1} n(ds,dv) \right). \tag{134}$$

Consider the two independent Poisson random measures

$$N_{\varepsilon,1}(ds,dv) = \mathbf{1}_{|v|<\varepsilon} N(ds,dv) \text{ and } N_{\varepsilon,2}(ds,dv) = \mathbf{1}_{|v|\geq\varepsilon} N(ds,dv).$$

Let $\Lambda_{\varepsilon,i}$ ($i = 1,2$) be the infinitely divisible random measure associated with $N_{\varepsilon,i}$ by (117). Remark that $X_{\varepsilon,1}^f$ and $X_{\varepsilon,2}^f$ are independent and that

$$X_{\varepsilon,i}^f(t) = \int_{\mathbb{R}^d} f(t,s)\,\Lambda_{\varepsilon,i}(ds), \quad i = 1,2.$$

In addition, the control measure $\mu_{\varepsilon,2}(dv) = \mathbf{1}_{|v|\geq\varepsilon} \mu(dv)$ of $\Lambda_{\varepsilon,2}$ is finite and symmetric. Therefore $X_{\varepsilon,2}^f$ can be simulated as a generalized shot noise series (see Sect. 6.3). It remains to properly approximate $X_{\varepsilon,1}^f$. To this task, notice that the control measure $\mu_{\varepsilon,1}(dv) = \mathbf{1}_{|v|<\varepsilon} \mu(dv)$ of $\Lambda_{\varepsilon,1}$ has moments of every order greater than 2. Hence, $\Lambda_{\varepsilon,1}$ is a Lévy random measure in the sense of [3].

Set

$$\sigma(\varepsilon) = \left(\int_{-\varepsilon}^{\varepsilon} v^2 \,\mu(dv) \right)^{1/2}. \tag{135}$$

Proposition 6.12. *Assume that for each* $t \in \mathbb{R}^d$, $f(t,\cdot) \in L^2(\mathbb{R}^d)$ *and* $\lim_{\varepsilon \to 0_+} \frac{\sigma(\varepsilon)}{\varepsilon} = +\infty$. *Then*

$$\lim_{\varepsilon \to 0_+} \left(\frac{X_{\varepsilon,1}^f(t)}{\sigma(\varepsilon)} \right)_{t \in \mathbb{R}^d} \overset{(d)}{=} \left(W^f(t) \right)_{t \in \mathbb{R}^d}, \tag{136}$$

where, with $W(ds)$ a real Brownian random measure defined in [9],

$$W^f(t) = \int_{\mathbb{R}^d} f(t, s) W(ds), \tag{137}$$

and, where the limit is understood in the sense of finite dimensional distributions.

Proof. Let $r \geq 1$, $t = (t_1, \ldots, t_r) \in (\mathbb{R}^d)^r$ and $y = (y_1, \ldots, y_r) \in \mathbb{R}^r$. Then

$$\mathbb{E}\left[\exp\left(i \sum_{k=1}^{r} y_k \frac{X_{\varepsilon,1}^f(t_k)}{\sigma(\varepsilon)}\right)\right] = \exp(\Psi_\varepsilon(t, y))$$

with

$$\Psi_\varepsilon(t, y)$$
$$= \int_{\mathbb{R}^d \times \mathbb{R}} \left(\exp\left(\frac{ig(s, t, y)v}{\sigma(\varepsilon)}\right) - 1 - \frac{ig(s, t, y)v}{\sigma(\varepsilon)} \mathbf{1}_{|g(s,t,y)v| \leq \sigma(\varepsilon)}\right) ds\, \mu_{\varepsilon,1}(dv)$$

and $g(s, t, y) = \sum_{k=1}^{r} y_k f(t_k, s)$. Then, by the Fubini theorem,

$$\Psi_\varepsilon(t, y) = \int_{\mathbb{R}^d} I_\varepsilon(g(s, t, y))\, ds,$$

where for every $a \in \mathbb{R}$ $I_\varepsilon(a) = \int_{\mathbb{R}} \left(e^{i\frac{av}{\sigma(\varepsilon)}} - 1 - i\frac{av}{\sigma(\varepsilon)} \mathbf{1}_{|av| < \sigma(\varepsilon)}\right) \mathbf{1}_{|v| < \varepsilon}\, \mu(dv)$.
Since $\mu(dv)$ is a symmetric Lévy measure,

$$I_\varepsilon(a) = \int_{\mathbb{R}} \left(e^{i\frac{av}{\sigma(\varepsilon)}} - 1 - i\frac{av}{\sigma(\varepsilon)}\right) \mathbf{1}_{|v| < \varepsilon}\, \mu(dv).$$

As $\lim_{\varepsilon \to 0+} \sigma(\varepsilon)/\varepsilon = +\infty$, according to [2], $\lim_{\varepsilon \to 0+} I_\varepsilon(a) = -\frac{a^2}{2}$. Since moreover $|I_\varepsilon(a)| \leq \frac{a^2}{2}$, for every $a \in \mathbb{R}$, a dominated convergence argument yields

$$\lim_{\varepsilon \to 0+} \Psi_\varepsilon(t, y) = -\frac{1}{2} \int_{\mathbb{R}^d} \left|\sum_{k=1}^{r} y_k f(t_k, s)\right|^2 ds = -\frac{1}{2} \mathrm{Var}\left(\sum_{k=1}^{r} y_k W^f(t_k)\right).$$

Hence we get that $\frac{X_{\varepsilon,1}^f(t)}{\sigma(\varepsilon)}$ converges in distribution to a Gaussian random variable. Moreover, if we recall that $\mathrm{Var} W^f(t) = \int_{\mathbb{R}^d} f^2(t, s)\, ds$ for a real Brownian random measure (136) is proved. □

As in the case of rhmLfs, an estimate in terms of Berry–Esseen bounds on the rate of convergence stated in Proposition 6.12 may be given. The assumption of the following theorem only ensures the existence of the moment of order $(2 + \delta)$ for $X_{\varepsilon,1}^f(t)$.

Theorem 6.13. *Let $t \in \mathbb{R}^d$ and assume that f satisfies (120) and that*

$$f(t, \cdot) \in L^{2+\delta}(\mathbb{R}^d) \tag{138}$$

for some $\delta \in (0, 1]$. Then $\mathbb{E}\left(\left|X_{\varepsilon,1}^f(t)\right|^{2+\delta}\right) < +\infty$, and

$$\sup_{u \in \mathbb{R}} \left|\mathbb{P}\left(X_{\varepsilon,1}^f(t) \leq u\right) - \mathbb{P}\left(\sigma(\varepsilon)W^f(t) \leq u\right)\right| \leq A(t, \delta) \frac{m_{2+\delta}^{2+\delta}(\varepsilon)}{\sigma^{2+\delta}(\varepsilon)}, \tag{139}$$

where W^f is defined by (137) in Proposition 6.12, $m_{2+\delta}^{2+\delta}(\varepsilon) = \int_{-\varepsilon}^{\varepsilon} |v|^{2+\delta} \mu(dv)$, and

$$A(t, \delta) = \frac{A_\delta \int_{\mathbb{R}^d} |f(t, s)|^{2+\delta} \, ds}{3 \left(\pi \int_{\mathbb{R}^d} |f(t, s)|^2 \, ds\right)^{(2+\delta)/2}},$$

with

$$A_\delta = \begin{cases} 0.7975 & \text{if } \delta = 1 \\ 53.9018 & \text{if } 0 < \delta < 1. \end{cases}$$

Remark 6.14. Assume that f satisfies assumptions (120) and (138). Then, for every t, $f(t, \cdot) \in L^2(\mathbb{R}^d)$ and $\mathbb{E}\left(\left|X_{\varepsilon,1}^f(t)\right|^2\right) < +\infty$.

Proof. This proof is based on a generalization of Lemma 4.1 in [22].

Let $\mathscr{L}$ be the distribution of the infinitely divisible variable $X_{\varepsilon,1}^f(t)$. The Lévy Q measure of $\mathscr{L}$ is then the push-forward of $n_{\varepsilon,1}(ds, dv) = ds\mu_{\varepsilon,1}(dv)$ by the map $(s, v) \mapsto f(t, s)v$. Hence, for every $\gamma > 0$,

$$\int_{\mathbb{R}} |y|^\gamma Q(dy) = m_\gamma^\gamma(\varepsilon) \int_{\mathbb{R}} |f(t, s)|^\gamma \, ds$$

where $m_\gamma^\gamma(\varepsilon) = \int_{-\varepsilon}^{\varepsilon} |v|^\gamma \mu(dv)$. Note that $m_2^2(\varepsilon) = \sigma^2(\varepsilon)$. Then, since $f(t, \cdot) \in L^{2+\delta}(\mathbb{R}^d)$,

$$\int_{\mathbb{R}} |y|^{2+\delta} Q(dy) < +\infty.$$

Therefore, according to Theorem 25.3 in [35],

$$\int_{\mathbb{R}} |y|^{2+\delta} \mathscr{L}(dy) < +\infty \quad \text{i.e.} \quad \mathbb{E}\left(\left| X_{\varepsilon,1}^f(t) \right|^{2+\delta} \right) < +\infty.$$

As in the proof of Lemma 4.1 in [22], we then consider a Lévy process $(Z(x))_{x \geq 0}$ such that $Z(1) \overset{(d)}{=} X_{\varepsilon,1}^f(t)$. For each fixed $n \in \mathbb{N} \backslash \{0\}$,

$$Z(1) = \sum_{k=0}^{n-1} \left(Z\left(\frac{k+1}{n} \right) - Z\left(\frac{k}{n} \right) \right)$$

where $Y_{k,n} = Z\left(\frac{k+1}{n} \right) - Z\left(\frac{k}{n} \right)$, $0 \leq k \leq n-1$, are i.i.d real-valued centered random variables. Furthermore,

$$\mathbb{E}\left(|Y_{k,n}|^2 \right) = \frac{\mathbb{E}\left(|Z(1)|^2 \right)}{n} = \frac{\sigma^2(\varepsilon) \int_{\mathbb{R}^d} |f(t,s)|^2 \, ds}{n}$$

and since $Z(1) \in L^{2+\delta}$, $Y_{k,n} \in L^{2+\delta}$. Therefore, according to [28], there exists a constant A_δ such that, for every $n \in \mathbb{N} \backslash \{0\}$,

$$\sup_{x \in \mathbb{R}} \left| \mathbb{P}\left(\frac{Z(1)}{\sqrt{\mathbb{E}\left(|Z(1)|^2 \right)}} \leq x \right) - \mathbb{P}(W \leq x) \right| \leq \frac{n A_\delta \mathbb{E}\left(\left| Z\left(\frac{1}{n} \right) \right|^{2+\delta} \right)}{\mathbb{E}\left(|Z(1)|^2 \right)^{1+\delta/2}}$$

where W is a normal random variable with mean 0 and variance 1. When $\delta = 1$, the preceding inequality is the classical Berry–Esseen inequality and we can take $A_1 = 0.7975$. In [28], one find that $A_\delta = \max(8/3, 64A_1 + 1 + 14/(3\sqrt{2\pi})) = 53.9018$. Furthermore, it is straightforward that

$$\sup_{x \in \mathbb{R}} \left| \mathbb{P}\left(\frac{Z(1)}{\sqrt{\mathbb{E}\left(|Z(1)|^2 \right)}} \leq x \right) - \mathbb{P}(W \leq x) \right|$$

$$= \sup_{u \in \mathbb{R}} \left| \mathbb{P}\left(X_{\varepsilon,1}^f(t) \leq u \right) - \mathbb{P}(\sigma(\varepsilon) W^f(t) \leq u) \right|.$$

According to [33],

$$\lim_{n \to +\infty} n \mathbb{E}\left(\left| Z\left(\frac{1}{n} \right) \right|^{2+\delta} \right) = \int_{\mathbb{R}} |y|^{2+\delta} Q(dy),$$

which concludes the proof. $\square$

6.5   Summary

We now summarize the approximation scheme based on the preceding splitting. First we approximate $X_{\varepsilon,1}^f$ by the Gaussian field $\sigma(\varepsilon)W^f$. According to Sect. 6.3, an approximation of $X_{\varepsilon,2}^f$ may be given by

$$Y_{\varepsilon,N,2}^f(t) = \sum_{n=1}^N f\left(t, \left(\frac{T_n}{c_d\,\mu_{\varepsilon,2}(\mathbb{R})}\right)^{1/d} U_n\right) V_{\varepsilon,n}, \quad t \in \mathbb{R}^d,$$

where $(V_{\varepsilon,n})_n$ is a sequence of i.i.d. random variables with common law $\mu_{\varepsilon,2}(dv)/\mu_{\varepsilon,2}(\mathbb{R})$. Note that T_n, U_n and $V_{\varepsilon,n}$ are independent. Since $X_{\varepsilon,1}^f$ and $X_{\varepsilon,2}^f$ are independent, W^f is assumed to be independent of $(T_n, U_n, V_{\varepsilon,n})$. As a result, in the case where $\mu(\mathbb{R}) = +\infty$, under the assumptions of Proposition 6.12, an approximation of X^f is

$$Y_{\varepsilon,N}^f(t) = \sigma(\varepsilon)W^f(t) + \sum_{n=1}^N f\left(t, \left(\frac{T_n}{c_d\,\mu_{\varepsilon,2}(\mathbb{R})}\right)^{1/d} U_n\right) V_{\varepsilon,n}, \quad t \in \mathbb{R}^d.$$

The choice of the cutoff ε is an important point for the simulation. The starting point for this choice is clearly the upper bound in (139). However this bound depends both on the Lévy measure μ and on the integrand f. Hence it is rather specific to each example. Nevertheless we can refer to the discussion in p. 489 of [2] to get the flavor of the calibration of the cutoff.

6.6   Examples

This section illustrates with various examples the range of application of the preceding results. In all the following examples, $K \subset \mathbb{R}^d$ is a compact set and (125) is only fulfilled for $\|s\| \geq A$. Then, as noticed in Remark 6.9, we may split

$$Y_N^f = Y_N^g + Y_N^{f-g},$$

with $g(t, s) = f(t, s)\mathbf{1}_{\|s\|\geq A}$. Since, Y_N^g is in fact a finite sum (almost surely), the rate of convergence described below is actually the rate of convergence of Y_N^{f-g}.

6.6.1   Moving Average Fractional Lévy Fields

Let $H \in (0, 1)$ such that $H \neq d/2$. Suppose that

$$f_{H,2}(t, s) = \|t - s\|^{H-d/2} - \|s\|^{H-d/2}$$

and that for every $p \geq 2$, $\int_{\mathbb{R}} |v|^p \, \mu(dv) < +\infty$. Then, $X_{H,2} = X^{f_{H,2}}$ is a mafLf in the sense of section 4.1.

6.6.2 Case of Finite Control Measures

An approximation, in law, of the mafLf X_H is given by

$$Y_N^{f_{H,2}}(t) = \sum_{n=1}^{N} \left(\left\| t - \left(\frac{T_n}{c_d \mu(\mathbb{R})} \right)^{1/d} U_n \right\| \right)^{H-d/2}$$

Let $A = \max_K \|y\| + 1$, $t \in K$ and $\|s\| \geq A$. The mean value inequality leads to

$$|f_{H,2}(t,s)| \leq \left| H - \frac{d}{2} \right| (A-1) \sup_{0 < \theta < 1} \|s - \theta t\|^{H-d/2-1}.$$

Remark that $\|s - \theta t\| \geq \|s\| - \|t\| \geq \|s\|/A$. Therefore, since $H - d/2 - 1 < 0$, for every $t \in K$, for $\|s\| \geq A$,

$$|f_{H,2}(t,s)| \leq \frac{C}{\|s\|^{1-H+d/2}} \tag{140}$$

with $C = |H - d/2|(A-1)A^{1-H+d/2}$.

Let $\beta_1 = 1 - H + d/2$ and $g_{H,2}(t,s) = f_{H,2}(t,s) \mathbf{1}_{\|s\| \geq \max_K \|y\|+1}$. Note that $\beta_1 > d/2$ since $1 > H$. Then, the assumptions of Theorem 6.7 are satisfied with $r = 2$ and

$$\mathbb{E}\left(\left| Y_N^{g_{H,2}}(t) - Y^{g_{H,2}}(t) \right|^2 \right) \leq \frac{C(2,\beta_1) D(N,2,\beta_1)}{N^{2(1-H)/d}}$$

where $C(2,\beta_1)$ and $D(N,2,\beta_1)$ are defined by (128) and (127). Therefore, the mean square error converges at the rate $N^{(1-H)/d}$.

We now focus on the uniform convergence of $Y^{g_{H,2}}$. For every integer $q \geq 1$, by a Taylor expansion, one can prove that for every $t \in K$ and for $\|s\| \geq A$,

$$\left| f_{H,2}(t,s) - \sum_{j=1}^{q-1} \|s\|^{H-j-d/2} d_j(t, s/\|s\|) \right| \leq B_{q,A,H} \|s\|^{H-d/2-q}, \tag{141}$$

for some positive constant $B_{q,A,H}$ and where the $d'_j s$ are polynomial functions in t_i and u_i, $i = 1 \dots d$, $j = 1 \dots d$. Since the d_j's are polynomial functions, one can easily see that $g_{H,2}$ satisfies the assumption (131) taking $\beta_1 = 1 - H + d/2$ and $\beta_p = q - H + d/2$. Since (141) holds for every integer $q \geq 1$, by Theorem 6.10, $Y_N^{g_{H,2}}$ converges uniformly at the rate N^ε for every $\varepsilon \in (0, (1-H)/d)$. Let us now present one example (see Fig. 1) taking $\mu(dv) = (\delta_{-1} + \delta_1)/2$. In this example, we first simulate a realization of the random variables (T_n, U_n, V_n). Then, for these values

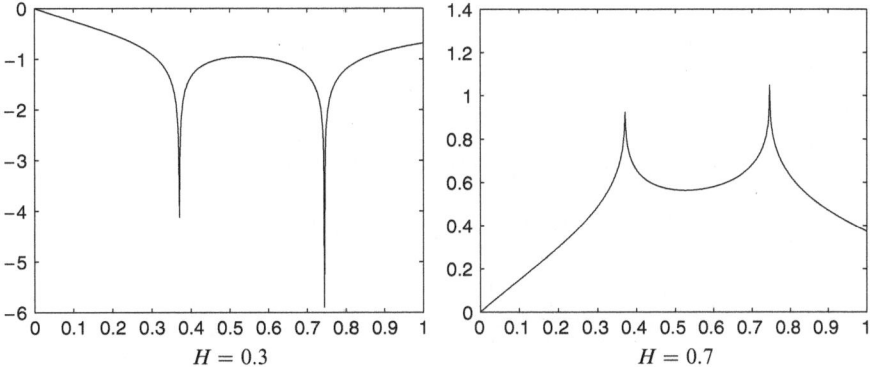

Fig. 1 Examples of mafLfs

of $(T_n, U_n, V_n)_{1 \leq n \leq N}$, we evaluate $Y_N^{f_H,2}$ for $H = 0.3$ and $H = 0.7$. We observe that the trajectory regularity does not depend on the value of H. Actually, one can see that the derivatives of $Y_N^{g_H,2}$, at each order, converge uniformly on each compact set. Therefore, $Y^{g_H,2}$ has $\mathscr{C}^\infty$ sample paths almost surely. As a consequence, the sample paths of $Y^{f_H,2}$ are $\mathscr{C}^\infty$ except at points $s_n = (T_n/c_d\mu(\mathbb{R}))^{1/d}U_n$. At these points, the behavior depends on H: while, when $H < d/2$, $Y_N^{f_H,2}$ is not defined, when $H > d/2$, the pointwise Hölder exponent of $Y_N^{f_H,2}$ is given by $H - d/2$. In Fig. 1, we observe that the sample paths are smooth on $[0, 1]$ except at two points.

6.6.3 Case of Infinite Control Measures

In this example,

$$\mu(dv) = \frac{\mathbf{1}_{|v| \leq 1}\, dv}{|v|^{1+\alpha}} \quad \text{with} \quad 0 < \alpha < 2.$$

Let $(V_{\varepsilon,n})_{n \geq 1}$ be a sequence of i.i.d variables with common law

$$\frac{\alpha\, \mathbf{1}_{\varepsilon < |v| < 1}\, dv}{2(\varepsilon^{-\alpha} - 1)|v|^{1+\alpha}}.$$

Moreover, let B_H be a standard fractional Brownian field (cf. Definition 3.1 with $C = 1$) with index H and assume that B_H, $(U_n)_{n \geq 1}$, $(T_n)_{n \geq 1}$ and $(V_{\varepsilon,n})_{n \geq 1}$ are independent. An approximation of the mafLf X_H is thus given by

$$Y_{\varepsilon,N}^{f_H,2}(t) = \sum_{n=1}^{N}\left(\left\|t - \left(\frac{T_n}{c_d\mu_{\varepsilon,2}(\mathbb{R})}\right)^{1/d}U_n\right\|^{H-d/2} - \left(\frac{T_n}{c_d\mu_{\varepsilon,2}(\mathbb{R})}\right)^{H/d-1/2}\right)V_{\varepsilon,n}$$

$$+ \sigma(\varepsilon)W^{f_H,2}(t), \qquad (142)$$

where

$$W^{f_{H,2}}(\cdot) = \int_{\mathbb{R}^d} f_{H,2}(\cdot, s) \, W(ds) \stackrel{(d)}{=} C_{H,d} B_H(\cdot),$$

with

$$C_{H,d} = \left(\int_{\mathbb{R}^d} |f_{H,2}(e_1, s)|^2 \, ds \right)^{1/2},$$

and $e_1 = (1, 0, \dots, 0)$. Actually, by a Fourier transform argument,

$$C_{H,d} = \frac{2^{H-2}|d - 2H|\Gamma(H/2 + d/4)}{\Gamma(d/4 + 1 - H/2)} \left(\int_{\mathbb{R}^d} \frac{|e^{-ie_1 \cdot \lambda} - 1|^2}{\|\lambda\|^{2H+d}} \, d\lambda \right)^{1/2}.$$

As a result, due to [34] for $d = 1$ and to [22] for $d \geq 2$,

$$C_{H,d} = \frac{2^{H-2}|d - 2H|\Gamma(H/2 + d/4)}{\Gamma(d/4 + 1 - H/2)} \left(\frac{\pi^{(d+1)/2}\Gamma(H + 1/2)}{H\Gamma(2H)\sin(\pi H)\Gamma(H + d/2)} \right)^{1/2}.$$

$$(143)$$

Since $H > 0$, there exists $\delta \in (0, 1]$ such that $H > d/2 - d/(2 + \delta)$, which implies that $f_{H,2}(t, \cdot) \in L^{2+\delta}(\mathbb{R}^d)$. Then, by Theorem 6.13, in terms of Berry–Esseen bounds, the rate of convergence of the error due to the approximation of $X_{\varepsilon,1}^f(t)$ is of the order

$$\delta(\varepsilon) = \frac{(2 - \alpha)^{1+\delta/2} \varepsilon^{\alpha\delta/2}}{(2 + \delta - \alpha)2^{\delta/2}}.$$

Except at points $s_n = (T_n/c_d\mu_{\varepsilon,2}(\mathbb{R}))^{1/d}U_n$, the trajectory regularity of $Y_{\varepsilon,N}^{f_{H,2}}$ is given by the trajectory regularity of $W^{f_{H,2}}$. Between two points s_n, the pointwise Hölder exponent of $Y_{\varepsilon,N}^{f_{H,2}}$ is equal to H. When $H > d/2$, the trajectories of $Y_{\varepsilon,N}^{f_{H,2}}$ are thus H'-Hölder on each compact set for every $H' < H - d/2$. Following [3], this is exactly what we expect for the trajectory regularity of a mafLf X_H associated to an infinite control measure. Figure 2 yields illustration of these facts in the case where $H = 0.8$, $\alpha = 1$, $d = 1$ and for the preceding control measure.

6.6.4 Moving Average Fractional Stable Fields

In this example,

$$\mu(dv) = \frac{dv}{|v|^{1+\alpha}} \text{ with } 0 < \alpha < 2, \qquad (144)$$

and

$$f_{H,\alpha}(t, s) = \left(\|t - s\|^{H-d/\alpha} - \|s\|^{H-d/\alpha} \right),$$

Fig. 2 Example of mafLf
with index $H = 0.8$

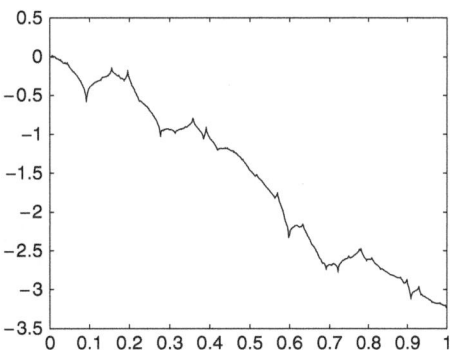

with $0 < H < 1$ and $H \neq d/\alpha$. Here $\sigma^2(\varepsilon) = 2\varepsilon^{2-\alpha}/(2 - \alpha)$ and $\mu_{\varepsilon,2}(\mathbb{R}) = 2/(\alpha\varepsilon^\alpha)$. The approximation of the mafsf is given by formula (142), replacing $d/2$ by d/α in the summation and with

$$W^{f_{H,\alpha}}(\cdot) = \int_{\mathbb{R}^d} f_{H,\alpha}(\cdot, s)\, W(ds) \overset{(d)}{=} C_{H+d/2-d/\alpha,d} B_{H+d/2-d/\alpha}(\cdot).$$

More precisely, as previously, $B_{H+d/2-d/\alpha}$ is a standard fBf with index $H + d/2 - d/\alpha$ and $C_{H+d/2-d/\alpha,d}$ is defined by (143). Furthermore, $\mu_{\varepsilon,2}(dv) = \mathbf{1}_{|v|>\varepsilon}\, \mu(dv)$ and $(V_{\varepsilon,n})_{n\geq 1}$, is a sequence of i.i.d variables with common law $\mu_{\varepsilon,2}(dv)/\mu_{\varepsilon,2}(\mathbb{R})$. Let us recall that the sequences B_H, $(U_n)_{n\geq 1}$, $(T_n)_{n\geq 1}$ and $(V_{\varepsilon,n})_{n\geq 1}$ are independent. Thus, the approximation of the mafsf $X_{H,\alpha} = X^{f_{H,\alpha}}$ is given by

$$Y_{\varepsilon,N}^{f_{H,\alpha}}(t) = \sum_{n=1}^{N} \left(\left\| t - \left(\frac{T_n}{c_d \mu_{\varepsilon,2}(\mathbb{R})} \right)^{1/d} U_n \right\|^{H-d/\alpha} - \left(\frac{T_n}{c_d \mu_{\varepsilon,2}(\mathbb{R})} \right)^{H/d-1/\alpha} \right) V_{\varepsilon,n}$$

$$+ \sigma(\varepsilon) W^{f_{H,\alpha}}(t).$$

However, this approximation only holds if $f_{H,\alpha}(t,\cdot) \in L^2(\mathbb{R}^d)$, i.e. the fBf $B_{H+d/2-d/\alpha}$ is defined, that is if $1 > H > d/\alpha - d/2$.

Observe that the asymptotic expansion of $f_{H,\alpha}$ is given by (141), replacing $d/2$ by d/α. Then, let $g_{H,\alpha}(t,s) = f_{H,\alpha}(t,s)\mathbf{1}_{\|s\|\geq \max_K \|y\|+1}$ and note that $Y_{\varepsilon,N}^{f_{H,\alpha}} = Y_{\varepsilon,N,2}^{g_{H,\alpha}} + Y_{\varepsilon,N,2}^{f_{H,\alpha}-g_{H,\alpha}} + \sigma(\varepsilon)W^{f_{H,\alpha}}(t)$ with

$$Y_{\varepsilon,N,2}^{h}(t) = \sum_{n=1}^{N} h\left(t, \frac{T_n}{c_d \mu_{\varepsilon,2}(\mathbb{R})} \right) V_{\varepsilon,N}.$$

Fig. 3 Example of mafsf
with index $H = 0.7$

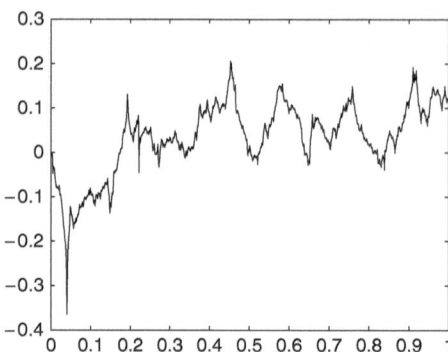

As noticed in Remark 6.9, $Y_{\varepsilon,N,2}^{f_{H,\alpha}-g_{H,\alpha}}$ is a finite sum. In addition, $g_{H,\alpha}$ satisfies the
assumptions of Theorem 6.7 for every $r < \alpha$. In this case therefore,

$$\mathbb{E}\left(\left|Y_{\varepsilon,N,2}^{g_{H,\alpha}}(t) - Y_{\varepsilon,2}^{g_{H,\alpha}}(t)\right|^r\right) \leq \frac{C(r,\beta)D(N,r,\beta)}{N^{r(1/d+1/\alpha-H/d)-1}},$$

where $\beta = 1+d/\alpha-H$. Furthermore, by Theorem 6.10, $Y_{\varepsilon,N,2}^{g_{H,\alpha}}$ converges uniformly
at the rate N^ε for every $\varepsilon \in (0,(1-H)/d)$.

Finally, when $H > d/\alpha - d/2$, there exists $\delta \in (0,1]$ such that $H > d/\alpha - d/(2+\delta)$. Then, $\mathbb{E}\left(\left|X_{\varepsilon,1}^f(t)\right|^{2+\delta}\right) < +\infty$ and as in the case of mafLfs, in terms of
Berry–Esseen bounds, the rate of convergence of the error due to the approximation
of $X_{\varepsilon,1}^f(t)$ is of the order

$$\delta(\varepsilon) = \frac{(2-\alpha)^{1+\delta/2}\varepsilon^{\alpha\delta/2}}{(2+\delta-\alpha)2^{\delta/2}}.$$

Except at points $s_n = (T_n/c_d\mu_{\varepsilon,2}(\mathbb{R}))^{1/d}U_n$, the pointwise Hölder exponent of
$Y_{\varepsilon,N,2}^{f_{H,\alpha}}$ is given by the one of $W^{f_{H,\alpha}}$ and thus is equal to $H - d/\alpha + d/2$. When
$H > d/\alpha$, on each compact set, $Y_{\varepsilon,N}^{f_{H,\alpha}}$ has H'-Hölder sample paths for every $H' < H - d/\alpha$. Figure 3 presents a realization of a mafsf when $\alpha = 1.5$ and $H = 0.7$.

6.6.5 Linear Fractional Stable Motions

Here $d = 1$ and we use the notation of Sect. 6.6.4. In particular, $\mu(dv)$ is given
by (144). In this example, the kernel function is

$$f(t,s) = \left((t-s)_+^{H-1/\alpha} - (-s)_+^{H-1/\alpha}\right),$$

where $(a)_+ = a \vee 0$, $H \in (0,1)$, $H \neq 1/\alpha$ (with the convention $0^{H-1/\alpha} = 0$). Hence, $L_{H,\alpha} = X^f$ is a linear fractional stable motion with index H (see [34] for more details on this process). Furthermore, we may approximate $L_{H,\alpha}$ in distribution by

$$Y^f_{\varepsilon,N}(t) = \sum_{n=1}^{N} \left(\left(t - \frac{T_n U_n}{2\mu_{\varepsilon,2}(\mathbb{R})} \right)_+^{H-1/\alpha} - \left(\frac{-T_n U_n}{2\mu_{\varepsilon,2}(\mathbb{R})} \right)_+^{H-1/\alpha} \right) V_{\varepsilon,n}$$
$$+ \sigma(\varepsilon) W^f(t),$$

where W^f is defined by (137). As previously, W^f is independent of $((U_n, T_n, V_{\varepsilon,n}))_{n \geq 1}$. Moreover,

$$W^f(\cdot) = \int_{\mathbb{R}^d} f(\cdot, s) \, W(ds) \overset{(d)}{=} \widetilde{C_H} B_{H+1/2-1/\alpha}$$

where $B_{H+1/2-1/\alpha}$ is a fBf with index $H + 1/2 - 1/\alpha$ and

$$\widetilde{C_H} = \left(\int_{\mathbb{R}} \left((t-s)_+^{H-1/\alpha} - (-s)_+^{H-1/\alpha} \right)^2 ds \right)^{1/2}$$

$$= \Gamma(H + 1/2) \sqrt{\frac{\sin\left((H - 1/\alpha)\pi\right)\Gamma(1 - 2H + 2/\alpha)}{2\pi(H + 1/2 - 1/\alpha)(H - 1/\alpha)}}$$

(145)

according to Lemma 4.1 in [40]. Obviously, this approximation only holds when $1 > H > 1/\alpha - 1/2$.

Furthermore, let us observe that

$$f(t,s) = \begin{cases} 0 & \text{if } s > \max_K |y| \\ f_{H,\alpha}(t,s) & \text{if } s < -\max_K |y|. \end{cases}$$

As a consequence, we obtain the same estimates for the almost sure, the L^r errors ($r < \alpha$), and the rate of convergence in terms of Berry–Esseen bounds, as in the case of mafsfs (see Sect. 6.6.4).

Figure 4 presents two realizations of lfsms for $\alpha = 1.5$. As noticed in [37], when $H = 0.2$, we observe spikes which take place at points s_n. Actually, since $H = 0.2 < 1/\alpha$, when t tends to a point s_n, $Y^f_{\varepsilon,N}(t)$ tends to infinity, which explains that spikes appear. When $H = 0.7 > 1/\alpha$, as in the case of mafsfs, the sample paths of the approximation are H'-Hölder on each compact set for every $H' < H - 1/\alpha$.

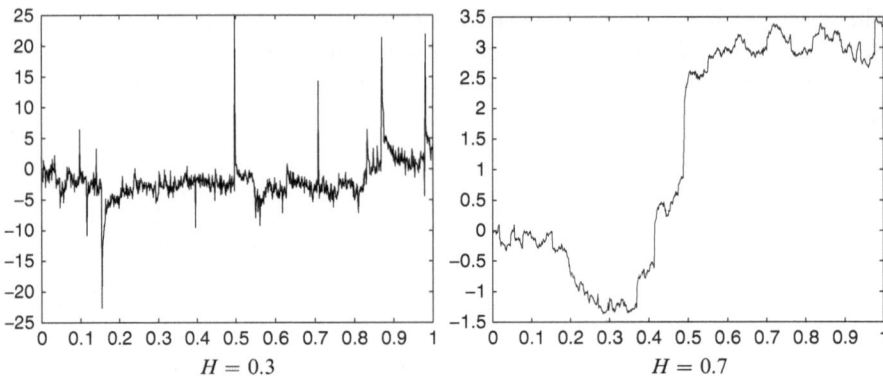

Fig. 4 Examples of Lfsms

6.6.6 Log-Fractional Stable Motion

Let $d = 1$ and $1 < \alpha < 2$, and assume that $\mu(dv)$ is given by (144). Furthermore, $(V_{\varepsilon,n})_{n \geq 1}$ and $\sigma(\varepsilon)$ are defined as in Sect. 6.6.4. Remark that here $(U_n)_{n \geq 1}$ is a sequence of i.i.d symmetric Bernoulli random variables. Then, let

$$f(t, s) = (\ln|t - s| - \ln|s|).$$

Hence, X^f is a log-fractional stable motion and its approximation in law is given by

$$Y^f_{\varepsilon,N}(t) = \sum_{n=1}^{N} \left(\ln\left| t - \frac{T_n U_n}{2\mu_{\varepsilon,2}(\mathbb{R})} \right| - \ln\left(\frac{T_n}{2\mu_{\varepsilon,2}(\mathbb{R})} \right) \right) V_{\varepsilon,n} + \sigma(\varepsilon) W^f(t),$$

where

$$W^f(t) = \int_{\mathbb{R}} (\ln|t - s| - \ln|s|) W(ds)$$

is independent of $((U_n, T_n, V_{\varepsilon,n}))_{n \geq 1}$. Note that $W^f \overset{(d)}{=} C B_{1/2}$, where $B_{1/2}$ is a standard Brownian motion and

$$C = \int_{\mathbb{R}} (\ln|1 - s| - \ln|s|)^2 \, ds.$$

Furthermore, by a Fourier transform argument, one proves that [34]

$$C = \left(\frac{\pi}{2} \int_{\mathbb{R}} \frac{\left| e^{-i\lambda} - 1 \right|^2}{|\lambda|^2} \, d\lambda \right)^{1/2} = \pi.$$

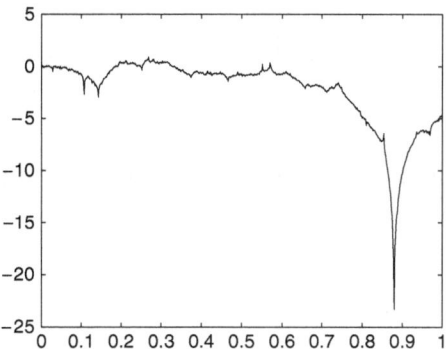

Fig. 5 Example of a log-fractional stable motion

As previously, the rate of almost sure convergence can be studied. In particular, if

$$g(t,s) = f(t,s)\mathbf{1}_{|s| \geq \max_K |y|+1},$$

$Y_{\varepsilon,N,2}^g$ converges uniformly on K at least at the rate N^ε for every $\varepsilon \in (0, 1 - 1/\alpha)$. Furthermore, the L^r-error can be controlled and decreases in $N^{1-1/r}$ for every $r < \alpha$. Let us notice that X^f is a self-similar field with index $H = 1/\alpha$. Thus, we obtain the same rate of convergence for log-fractional stable motion and mafsfs. Furthermore, since $f(t, \cdot) \in L^3(\mathbb{R})$, Theorem 6.13 gives the same rate of convergence in terms of Berry–Esseen bounds as in the cases of mafsfs or mafLfs (taking $\delta = 1$).

Figure 5 presents a trajectory of a log-fractional stable motion for $\alpha = 1.5$. Note that except at points $s_n = T_n U_n/(2\mu_{\varepsilon,2}(\mathbb{R}))$, the sample paths are locally H'-Hölder for every $H' < 1/2$: actually the regularity of the trajectories is given by the Brownian part. At points s_n, $Y_{\varepsilon,N}^f$ is not defined, which explains that spikes appear in Fig. 5.

6.6.7 Linear Multifractional Stable Motions

So far, the examples are fractional fields. However, our framework also contains multifractional fields. Let us now give one example. This example is defined replacing in the kernel of a LFSM the index H by $h(t)$.

Here $d = 1$ and $\mu(dv)$ is given by (144). Then, assume that the kernel function is defined by

$$f(t,s) = (t-s)_+^{h(t)-1/\alpha} - (-s)_+^{h(t)-1/\alpha}$$

where $h : \mathbb{R} \to (0, 1)$. The process X^f is a linear multifractional stable motion in the sense of [38, 39]. The approximation of X^f is then given by

$$Y_{\varepsilon,N}^f(t) = \sum_{n=1}^N \left(\left(t - \frac{T_n U_n}{2\mu_{\varepsilon,2}(\mathbb{R})} \right)_+^{h(t)-1/\alpha} - \left(\frac{-T_n U_n}{2\mu_{\varepsilon,2}(\mathbb{R})} \right)_+^{h(t)-1/\alpha} \right) V_{\varepsilon,n}$$
$$+ \sigma(\varepsilon) W^f(t),$$

where W^f is defined by (137). As previously, W^f is independent of $((U_n, T_n, V_{\varepsilon,n}))_{n\geq 1}$. Moreover,

$$W^f(\cdot) = \int_{\mathbb{R}^d} f(\cdot, s) \, W(ds) \stackrel{(d)}{=} \widetilde{C_{h(t)}} B_{h+1/2-1/\alpha}$$

where $B_{h+1/2-1/\alpha}$ is a standard multifractional Brownian motion in the sense of [27] with multifractional function $h + 1/2 - 1/\alpha$ and $\widetilde{C_{h(t)}}$ is given by (145). This approximation only holds when $1 > h(t) > 1/\alpha - 1/2$.

As in the case of Lfsm, we can observe that

$$f(t, s) = \begin{cases} 0 & \text{if } s > \max_K |y| \\ f_{h(t),\alpha}(t, s) & \text{if } s < -\max |y|. \end{cases}$$

Therefore, for a fixed t, we obtain the same estimates for the almost sure, the L^r errors ($r < \alpha$), and the rate of convergence in terms of Berry–Esseen bounds, as in the case of Lfsm (see Sect. 6.6.5) or mafsfs (see Sect. 6.6.4), replacing H by $h(t)$. In particular, for a fixed t, the almost sure error converges at the rate N^ε for every $\varepsilon \in (0, 1 - h(t))$.

Figure 6 presents some trajectories of linear multifractional stable motions for $\alpha = 1.5$.

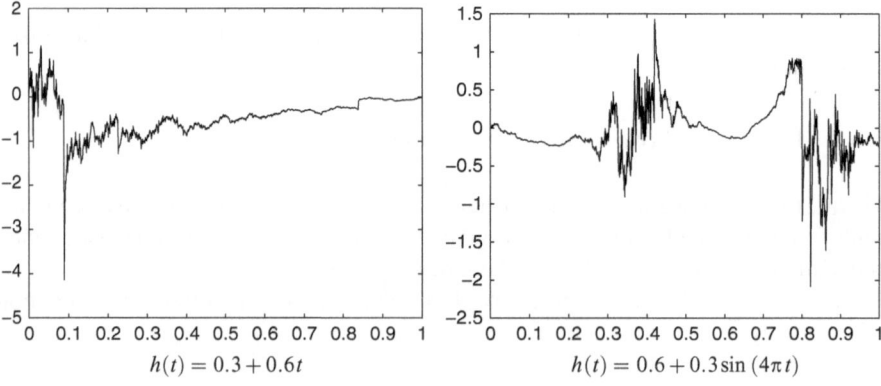

$$h(t) = 0.3 + 0.6t \qquad\qquad h(t) = 0.6 + 0.3\sin(4\pi t)$$

Fig. 6 Examples of linear multifractional stable motions

6.7 Simulation of Harmonizable Fields

In the Sects. 4.2 and 4.4 we have studied fractional or multifractional Lévy fields, which are integrals of complex random measures. Thanks to arguments used in Sects. 6.3 and 6.4, the results obtained in the case of rhmLfs in [22] can be extended to a larger class of infinitely divisible fields, in particular to the complex case. Let us recall that Poisson random measures on $\mathbb{R}^d \times \mathbb{C}$ are denoted by $N(d\xi, dz)$ with intensity $n(d\xi, dz) = d\xi \mu(dz)$. Here in contrast to (16) we assume that $\mu(dz)$ is a σ-finite measure such that

$$\int_{\mathbb{C}} \left(|z|^2 \wedge 1 \right) \mu(dz) < +\infty.$$

The control measure $\mu(dz)$ is assumed to be rotationally invariant as in (15)

$$P(\mu(dz)) = d\theta \, \mu_\rho(d\rho),$$

where $d\theta$ is the uniform measure on $[0, 2\pi)$ and $P\left(\rho e^{i\theta}\right) = (\theta, \rho) \in [0, 2\pi) \times \mathbb{R}_*^+$.

Then, following the definition of complex Lévy random measure (see [7]), we can consider a complex random measure $\Lambda(d\xi)$ on $\mathbb{R}^d$ defined by

$$\int_{\mathbb{R}^d} g(\xi) \, \Lambda(d\xi) = \int_{\mathbb{R}^d \times \mathbb{C}} (g(\xi) z + g(-\xi) \overline{z})$$

$$\times \left(N(d\xi, dz) - (|g(\xi) z + g(-\xi) \overline{z}| \vee 1)^{-1} n(d\xi, dz) \right) \quad (146)$$

for every $g : \mathbb{R}^d \to \mathbb{C}$ such that $\int_{\mathbb{R}^d \times \mathbb{C}} \left(|g(\xi) z|^2 \wedge 1 \right) d\xi \mu(dz) < +\infty$.

Hence, following the arguments used in [22] in the case of rhfLfs, analogous results to those of Sects. 6.3 and 6.4 can be obtained and a way to simulate

$$X^f(x) = \int_{\mathbb{R}^d} f(x, \xi) \, \Lambda(d\xi).$$

can be proposed. However, in this part, we will just focus on the case, where

$$\mu_\rho(d\rho) = \frac{\mathbf{1}_{\rho > 0} \, d\rho}{\rho^{1+\alpha}}, \quad \alpha \in (0, 2),$$

and the kernel function is

$$f_{H,\alpha}(x, \xi) = \frac{\left(2^{\alpha+1} \pi \right)^{-1/\alpha} \left(e^{-ix \cdot \xi} - 1 \right)}{\|\xi\|^{H+d/\alpha}}.$$

In this case,

$$X_{H,\alpha}(x) = \int_{\mathbb{R}^d} f_{H,\alpha}(x,\xi)\, \Lambda(d\xi), \quad x \in \mathbb{R}^d,$$

is a real harmonizable fractional stable field with index $H \in (0,1)$ of Sect. 3.2.2, i.e.

$$\{X_{H,\alpha}(x),\, x \in \mathbb{R}^d\} \overset{(d)}{=} \{S_{H,\alpha}(x),\, x \in \mathbb{R}^d\},$$

where

$$S_H(x) = \int_{\mathbb{R}^d} f_{H,\alpha}(x,\xi)\, M_\alpha(d\xi),$$

with $M_\alpha(d\xi)$ a complex isotropic α-stable random measure in the sense of Definition 2.8. Furthermore, in the case we are interested in, $\mu(\mathbb{C}) = +\infty$. As we know, we have to split in this case the random field $X_{H,\alpha} = X_{\varepsilon,1}^{f_{H,\alpha}} + X_{\varepsilon,2}^{f_{H,\alpha}}$ into two random fields, where

$$X_{\varepsilon,1}^{f_{H,\alpha}}(x) = 2 \int_{\mathbb{R}^d \times \mathbb{C}} \Re(f_{H,\alpha}(x,\xi)z)\mathbf{1}_{|z|<\varepsilon}$$
$$\times \left(N(d\xi, dz) - (|2\Re(f_{H,\alpha}(x,\xi)z)| \vee 1)^{-1}\, n(d\xi, dz) \right), \quad (147)$$

and

$$X_{\varepsilon,2}^{f_{H,\alpha}}(x) = 2 \int_{\mathbb{R}^d \times \mathbb{C}} \Re(f_{H,\alpha}(x,\xi)z)\mathbf{1}_{|z|\geq\varepsilon}$$
$$\times \left(N(d\xi, dz) - (|2\Re(f_{H,\alpha}(x,\xi)z)| \vee 1)^{-1}\, n(d\xi, dz) \right). \quad (148)$$

As previously, $X_{\varepsilon,1}^{f_{H,\alpha}}$ and $X_{\varepsilon,2}^{f_{H,\alpha}}$ can be simulated independently. Furthermore,

$$X_{\varepsilon,2}^{f}(x) = \int_{\mathbb{R}^d} f_{H,\alpha}(x,\xi)\, \Lambda_{\varepsilon,2}(d\xi), \quad x \in \mathbb{R}^d,$$

where the complex random measure $\Lambda_{\varepsilon,2}$ is associated by (146) to a Poisson random measure $N_{\varepsilon,2}$, whose control measure $\mu_{\varepsilon,2}(dz) = \mathbf{1}_{|z|\geq\varepsilon}\,\mu(dz)$ is finite. Therefore, $X_{\varepsilon,2}^{f}$ can be simulated as a generalized shot noise series. More precisely, let $(Z_{\varepsilon,n})_{n\geq 1}$ be a sequence of i.i.d. random variables with common law $\mu_{\varepsilon,2}(dz)/\mu_{\varepsilon,2}(\mathbb{C})$. Moreover, $(Z_{\varepsilon,n})_{n\geq 1}$, $(T_n)_{n\geq 1}$ and $(U_n)_{n\geq 1}$ are independent. Then, as in the case of rhmLfs, a series expansion of $X_{\varepsilon,2}^{f}$ can be given and these series converge in the space of continuous functions endowed with the topology of the uniform convergence on compact sets.

Proposition 6.15. *For every $x \in \mathbb{R}^d$,*

$$Y_{\varepsilon,N}^{f_{H,\alpha}}(x) = 2 \sum_{n=1}^{N} \Re\left(f_{H,\alpha}\left(x, \left(\frac{T_n}{c_d\,\mu(\mathbb{C})}\right)^{1/d} U_n\right) Z_{\varepsilon,n}\right) \qquad (149)$$

converges almost surely to $Y_\varepsilon^{f_{H,\alpha}}(x)$ as $N \to +\infty$. Furthermore, $Y_{\varepsilon,N}^{f_{H,\alpha}}$ converges uniformly on each compact set almost surely and

$$\{X_{\varepsilon,2}^{f_{H,\alpha}}(x) : x \in \mathbb{R}^d\} \overset{(d)}{=} \{Y_\varepsilon^{f_{H,\alpha}}(x) : x \in \mathbb{R}^d\}.$$

Proof. The arguments of proof of Proposition 6.5 lead to the almost sure convergence of $Y_{\varepsilon,N}^{f_{H,\alpha}}(x)$ for each fixed x. They also give the equality of the finite dimensional marginals of $X_{\varepsilon,2}^{f_{H,\alpha}}$ and $Y_\varepsilon^{f_{H,\alpha}}$. In order to obtain the uniform convergence, one may follow the proof of Proposition 3.1 in [22]. $\qquad\square$

Due to the rotational invariance of $Z_{\varepsilon,n}$ and to Theorem 6.7, a rate of almost sure convergence for $Y_{\varepsilon,N}^{f_{H,\alpha}}(x)$ can be given and the L^r-error can be controlled.

Proposition 6.16. *Let $x \in \mathbb{R}^d$.*

1. Then, for every $\varepsilon \in (0, H/d)$, almost surely,

$$\sup_{N \geq 1} N^\varepsilon \left| Y_\varepsilon^{f_{H,\alpha}}(x) - Y_{\varepsilon,N}^{f_{H,\alpha}}(x) \right| < +\infty.$$

2. Moreover, for every $r < \alpha$ and every integer $N > r(1/\alpha + H/d)$,

$$\mathbb{E}\left(\left| Y_{\varepsilon,N}^{f_{H,\alpha}}(x) - Y_\varepsilon^{f_{H,\alpha}}(x) \right|^r\right) \leq C(r)\,\frac{D(N, r, H + d/\alpha)}{N^{r/\alpha + rH/d - 1}}, \qquad (150)$$

where $D(N, r, \beta)$ is defined by (127) and

$$C(r) = \frac{\left(2^{1-\alpha}\pi\right)^{-r/\alpha} d(c_d\mu(\mathbb{R}))^{rH/d + r/\alpha}\mathbb{E}(|\Re(V_1)|^r)}{rH - d + rd/\alpha}.$$

Proof. Since $(Z_{\varepsilon,n})_{n \geq 1}$ is a sequence of i.i.d. with common law invariant by rotation,

$$Y_{\varepsilon,N}^{f_{H,\alpha}}(x) \overset{(d)}{=} 2 \sum_{n=1}^{N} \left| f_{H,\alpha}\left(x, \left(\frac{T_n}{c_d\,\mu(\mathbb{C})}\right)^{1/d} U_n\right) \right| \Re(Z_{\varepsilon,n}).$$

Hence, taking $V_n = \Re(Z_{\varepsilon,n})$, $C = 2^{1-1/\alpha}(\pi)^{-1/\alpha}$ and $\beta = H + d/\alpha$, the proof of Theorem 6.7 leads to the conclusion. $\qquad\square$

Finally, the next proposition gives the expected approximation of $X_{\varepsilon,1}$. Let

$$\sigma(\varepsilon) = \left(\int_0^\varepsilon \rho^2 \, \mu_\rho(d\rho) \right)^{1/2} = \sqrt{\frac{2\varepsilon^{2-\alpha}}{2-\alpha}}. \qquad (151)$$

Proposition 6.17. *Assume that* $0 < H + d/\alpha - d/2 < 1$ *then*

$$\lim_{\varepsilon \to 0_+} \left(\frac{X_{\varepsilon,1}(x)}{\sigma(\varepsilon)} \right)_{x \in \mathbb{R}^d} \overset{(d)}{=} \left(\Lambda_{H+d/\alpha-d/2} B_{H+d/\alpha-d/2}(x) \right)_{x \in \mathbb{R}^d},$$

where the convergence is in distribution on the space of continuous functions endowed with the topology of uniform convergence on compact sets, $B_{H+d/\alpha-d/2}$ *is a standard fBf with index* $H + d/\alpha - d/2$ *and for* $u \in (0,1)$

$$\Lambda_u = \left(2^{\alpha+1} \pi \right)^{-1/\alpha} \left(\frac{4 \pi^{(d+3)/2} \, \Gamma(u+1/2)}{u \, \Gamma(2u) \sin(\pi u) \, \Gamma(u+d/2)} \right)^{1/2}. \qquad (152)$$

Remark 6.18. In Proposition 6.17, $0 < H + d/\alpha - d/2 < 1$ means that $f_{H,\alpha}(x,\cdot) \in L^2(\mathbb{R}^d)$ for every $x \in \mathbb{R}^d$.

Proof. Actually

$$X_{\varepsilon,1}^f(x) = \int_{\mathbb{R}^d} f_{H,\alpha}(x,\xi) \, \Lambda_{\varepsilon,1}(d\xi), \quad x \in \mathbb{R}^d,$$

where the complex random measure $\Lambda_{\varepsilon,1}$ is associated by (146) with a Poisson random measure $N_{\varepsilon,1}$, whose control measure $\mu_{\varepsilon,1}(dz) = \mathbf{1}_{|z|<\varepsilon} \, \mu(dz)$. Also, for every $p \geq 2$,

$$\int_{\mathbb{C}} |z|^p \, \mu_{\varepsilon,1}(dz) < +\infty$$

and then $\left(2^{\alpha+1}\pi \right)^{1/\alpha} X_{\varepsilon,1}^f(x)$ is a rhfLf since $f_{H,\alpha}(x,\cdot) \in L^2(\mathbb{R}^d)$ for every $x \in \mathbb{R}^d$. Then, Proposition 4.1 in [22] yields the conclusion.   □

As a consequence, as soon as the assumptions of Proposition 6.17 are fulfilled, we may approximate the Rhfsf $X_{H,\alpha}$ by

$$Y_{\varepsilon,N}(x) = 2 \sum_{n=1}^N \Re \left(f_{H;\alpha} \left(x, \left(\frac{T_n}{c_d \, \mu(\mathbb{C})} \right)^{1/d} U_n \right) Z_{\varepsilon,n} \right)$$

$$+ \sigma(\varepsilon) \Lambda_{H+d/\alpha-d/2} B_{H+d/\alpha-d/2}(x), \quad x \in \mathbb{R}^d,$$

where $B_{H+d/\alpha-d/2}$, T_n, U_n and $Z_{\varepsilon,n}$ are independent.

Figure 7 exhibits some examples of trajectories of Rhfsms for $\alpha = 1.5$.

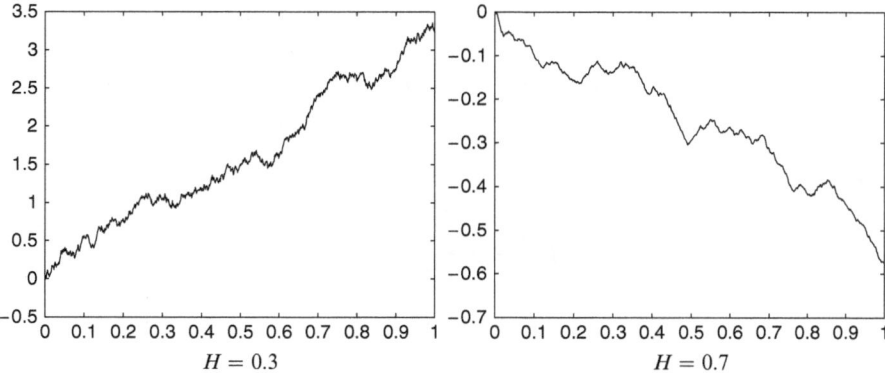

$H = 0.3$ $\qquad\qquad\qquad\qquad\qquad$ $H = 0.7$

Fig. 7 Examples of Rhfsfs

Appendix

Proof of Lemma 5.3

We first prove (108). Let us recall that

$$Z_{\tilde{H}}(t,t') = \frac{X_{H,\alpha}(t') - X_{H,\alpha}(t)}{\|t' - t\|^{\tilde{H}}}$$

for $t \neq t'$. Let us denote by $\epsilon = \|t - t'\|$ and by $\tau = \frac{t-t'}{\epsilon}$ a vector of Euclidean norm 1 that depends on t, t'. Since mafLm's have stationary increments, the characteristic function of $Z_{\tilde{H}}$ can be written

$$\mathbb{E} \exp\left(\frac{i\lambda X_{H,\alpha}(\epsilon\tau)}{\epsilon^{\tilde{H}}}\right)$$

$$= \exp\int_{\mathbb{R}^d} ds \int_{\mathbb{R}} \left[\exp\left(\frac{i\lambda u(\|\epsilon\tau - s\|^{\tilde{H}-d/\alpha} - \|s\|^{\tilde{H}-d/\alpha})}{\epsilon^{\tilde{H}}}\right) - 1\right.$$

$$\left. - \frac{i\lambda u(\|\epsilon\tau - s\|^{\tilde{H}-d/\alpha} - \|s\|^{\tilde{H}-d/\alpha})}{\epsilon^{\tilde{H}}}\right] \frac{\mathbf{1}(|u| \leq 1)du}{|u|^{1+\alpha}}.$$

$$\text{(A.1)}$$

Then the change of variable $s = \epsilon\sigma$ and $\epsilon^{-d/\alpha}u = w$ leads to

$$\mathbb{E} \exp\left(\frac{i\lambda X_{H,\alpha}(\epsilon\tau)}{\epsilon^{\tilde{H}}}\right) = \exp\int_{\mathbb{R}^d} d\sigma \int_{\mathbb{R}} [\exp\left(iw\lambda g(\sigma)\right) - 1 - iw\lambda g(\sigma)]$$

$$\times \frac{\mathbf{1}(|w| \leq \epsilon^{-d/\alpha})dw}{|w|^{1+\alpha}},$$

where $g(\sigma) = \|\tau - \sigma\|^{\tilde{H}-d/\alpha} - \|\sigma\|^{\tilde{H}-d/\alpha}$. Let us recall (12): $\forall \, 0 < \alpha < 2$

$$C_\alpha |x|^\alpha = \int_{\mathbb{R}} \frac{1 - e^{ixy} - ixy\mathbf{1}(|y| \le 1)}{|y|^{1+\alpha}} dy, \ \forall x \in \mathbb{R},$$

where

$$C_\alpha = \int_{\mathbb{R}} \frac{1 - e^{iy} - iy\mathbf{1}(|y| \le 1)}{|y|^{1+\alpha}} dy$$

$$= 2 \int_0^\infty \frac{1 - \cos(y)}{|y|^{1+\alpha}} dy > 0.$$

Let us define

$$H_\epsilon(x) = 2 \int_{\epsilon^{d/\alpha}}^\infty \frac{1 - \cos(yx)}{|y|^{1+\alpha}} dy. \tag{A.2}$$

Then

$$\forall x \in \mathbb{R}, H_\epsilon(x) \le 2 \int_{\epsilon^{d/\alpha}}^\infty \frac{dy}{|y|^{1+\alpha}} = \frac{2}{\alpha}\epsilon^d.$$

Moreover

$$H_\epsilon(x) = 2|x|^\alpha \int_{\epsilon^{d/\alpha}|x|}^\infty \frac{1 - \cos(v)}{|v|^{1+\alpha}} dv \le C_\alpha |x|^\alpha.$$

Hence

$$0 \le H_\epsilon(x) \le \inf(\frac{2}{\alpha}\epsilon^d, C_\alpha |x|^\alpha). \tag{A.3}$$

With the help of H_ϵ, one can rewrite

$$\int_{\mathbb{R}} [\exp(iw\lambda g(\sigma)) - 1 - iw\lambda g(\sigma)] \frac{\mathbf{1}(|w| \le \epsilon^{-d/\alpha})dw}{|w|^{1+\alpha}}$$

$$= \int_{\mathbb{R}} \left[\exp(iw\lambda g(\sigma)) - 1 - iw\lambda g(\sigma)\mathbf{1}(|w| \le \epsilon^{-d/\alpha})\right] \frac{dw}{|w|^{1+\alpha}}$$

$$+ \int_{\mathbb{R}} [\exp(iw\lambda g(\sigma)) - 1] \frac{\mathbf{1}(|w| > \epsilon^{-d/\alpha})dw}{|w|^{1+\alpha}}.$$

Hence

$$\int_{\mathbb{R}} [\exp(iw\lambda g(\sigma)) - 1 - iw\lambda g(\sigma)] \frac{\mathbf{1}(|w| \le \epsilon^{-d/\alpha})dw}{|w|^{1+\alpha}}$$

$$= -C_\alpha |\lambda g(\sigma)|^\alpha - H_\epsilon(\lambda g(\sigma)).$$

The last line is obtained using (12) and $\int_{\mathbb{R}} w 1(1 < |w| < \epsilon^{-d/\alpha}) \frac{dw}{|w|^{1+\alpha}} = 0$, for every $\epsilon < 1$. Inequality (A.3) implies that $\int_{\mathbb{R}^d} H_\epsilon(\lambda g(\sigma)) d\sigma < \infty$. Hence for $0 < \epsilon < 1$

$$\mathbb{E} \exp\left(\frac{i\lambda X_{H,\alpha}(\epsilon\tau)}{\epsilon^{\tilde{H}}}\right) = \exp\left(-C_\alpha |\lambda|^\alpha \int_{\mathbb{R}^d} |g(\sigma)|^\alpha d\sigma - \int_{\mathbb{R}^d} H_\epsilon(\lambda g(\sigma)) d\sigma\right). \tag{A.4}$$

Because of (A.4)

$$\mathbb{E} \exp\left(\frac{i\lambda X_{H,\alpha}(\epsilon\tau)}{\epsilon^{\tilde{H}}}\right) \leq \exp\left(-C_\alpha |\lambda|^\alpha \int_{\mathbb{R}^d} |g(\sigma)|^\alpha d\sigma\right)$$

and the function of λ on the right hand of the inequality is in $L^1(\mathbb{R}^d)$ for the Lebesgue measure $d\lambda$.

Let us now prove (109). Let us show that $\sup_{0<\epsilon<1} \mathbb{E} \left|\frac{X_{H,\alpha}(\epsilon\tau)}{\epsilon^{\tilde{H}}}\right|^\beta < +\infty$. Because of (12) for $0 < \beta < \alpha < 2$,

$$C_\beta \left|\frac{X_{H,\alpha}(\epsilon\tau)}{\epsilon^{\tilde{H}}}\right|^\beta$$
$$= \int_{\mathbb{R}} \left(1 - \exp\left(\frac{i\lambda X_{H,\alpha}(\epsilon\tau)}{\epsilon^{\tilde{H}}}\right) - \frac{i\lambda X_{H,\alpha}(\epsilon\tau)}{\epsilon^{\tilde{H}}} 1(|\lambda| \leq 1)\right) \frac{d\lambda}{|\lambda|^{1+\beta}}. \tag{A.5}$$

To compute the expectation of the left hand side of (A.5) we first check that we can apply Fubini theorem. Please remark that for every $M > 0$ there exists $0 < C_M < +\infty$ such that

$$\left|1 - \exp\left(\frac{i\lambda X_{H,\alpha}(\epsilon\tau)}{\epsilon^{\tilde{H}}}\right) - \frac{i\lambda X_{H,\alpha}(\epsilon\tau)}{\epsilon^{\tilde{H}}} 1(|\lambda| \leq 1)\right| \leq$$

$$C_M \left|\frac{\lambda X_{H,\alpha}(\epsilon\tau)}{\epsilon^{\tilde{H}}}\right|^2 1\left(\left|\frac{\lambda X_{H,\alpha}(\epsilon\tau)}{\epsilon^{\tilde{H}}}\right| \leq M\right)$$

$$+ \left(2 + \left|\frac{\lambda X_{H,\alpha}(\epsilon\tau)}{\epsilon^{\tilde{H}}}\right| 1(|\lambda| \leq 1)\right) 1\left(\left|\frac{\lambda X_{H,\alpha}(\epsilon\tau)}{\epsilon^{\tilde{H}}}\right| > M\right). \tag{A.6}$$

For the first line of the upper bound we get

$$\mathbb{E} \int_{\mathbb{R}} \left|\frac{\lambda X_{H,\alpha}(\epsilon\tau)}{\epsilon^{\tilde{H}}}\right|^2 1\left(\left|\frac{\lambda X_{H,\alpha}(\epsilon\tau)}{\epsilon^{\tilde{H}}}\right| \leq M\right) \frac{d\lambda}{|\lambda|^{1+\beta}} \leq C\mathbb{E}\left(\left|\frac{X_{H,\alpha}(\epsilon\tau)}{\epsilon^{\tilde{H}}}\right|^\beta\right),$$

and hence it is finite. Let us check the integrability of the other line in (A.6) Since $\mathbb{E}|X_{H,\alpha}(\epsilon\tau)|^2 < +\infty$, we have

$$\mathbb{P}\left(\left|\frac{\lambda X_{H,\alpha}(\epsilon\tau)}{\epsilon^{\tilde{H}}}\right| > M\right) \leq C\inf(1,|\lambda|^2).$$

Hence

$$\int_{\mathbb{R}} \mathbb{P}\left(\left|\frac{\lambda X_{H,\alpha}(\epsilon\tau)}{\epsilon^{\tilde{H}}}\right| > M\right)\frac{d\lambda}{|\lambda|^{1+\beta}} < \infty.$$

Then by Cauchy Schwartz inequality

$$\mathbb{E}\left(\left|\frac{\lambda X_{H,\alpha}(\epsilon\tau)}{\epsilon^{\tilde{H}}}\right|\mathbf{1}\left(\left|\frac{\lambda X_{H,\alpha}(\epsilon\tau)}{\epsilon^{\tilde{H}}}\right| > M\right)\right) \leq C\lambda^2$$

when $\lambda \to 0$. Hence

$$\int_{|\lambda|\leq 1} \mathbb{E}\left(\left|\frac{\lambda X_{H,\alpha}(\epsilon\tau)}{\epsilon^{\tilde{H}}}\right|\mathbf{1}\left(\left|\frac{\lambda X_{H,\alpha}(\epsilon\tau)}{\epsilon^{\tilde{H}}}\right| > M\right)\right)\frac{d\lambda}{|\lambda|^{1+\beta}} < \infty.$$

Consequently

$$\int_{\mathbb{R}} \mathbb{E}\left|1 - \exp\left(\frac{i\lambda X_{H,\alpha}(\epsilon\tau)}{\epsilon^{\tilde{H}}}\right) - \frac{i\lambda X_{H,\alpha}(\epsilon\tau)}{\epsilon^{\tilde{H}}}\mathbf{1}(|\lambda| \leq 1)\right|\frac{d\lambda}{|\lambda|^{1+\beta}} < \infty.$$

Then by Fubini theorem, and since $\mathbb{E}X_{H,\alpha}(\epsilon\tau) = 0$,

$$\mathbb{E}\left|\frac{X_{H,\alpha}(\epsilon\tau)}{\epsilon^{\tilde{H}}}\right|^\beta = \frac{1}{C_\beta}\int_{\mathbb{R}}\frac{1 - \mathbb{E}\exp\left(\frac{i\lambda X_{H,\alpha}(\epsilon\tau)}{\epsilon^{\tilde{H}}}\right)}{|\lambda|^{1+\beta}}d\lambda. \tag{A.7}$$

Let us remark that

$$\int_{|\lambda|>1}\frac{1 - \mathbb{E}\exp\left(\frac{i\lambda X_{H,\alpha}(\epsilon\tau)}{\epsilon^{\tilde{H}}}\right)}{|\lambda|^{1+\beta}}d\lambda < M,$$

where M is a finite and positive constant that does not depend on ϵ. Using (A.4) and $1 - e^{-x} \leq x$, for

$$x = C_\alpha|\lambda|^\alpha\int_{\mathbb{R}^d}|g(\sigma)|^\alpha d\sigma + \int_{\mathbb{R}^d}H_\epsilon(\lambda g(\sigma))d\sigma \geq 0,$$

we get

$$
\int_{|\lambda|\leq 1} \frac{1 - \mathbb{E}\exp\left(\frac{i\lambda X_{H,\alpha}(\epsilon\tau)}{\epsilon^{\tilde{H}}}\right)}{|\lambda|^{1+\beta}}\,d\lambda
$$

$$
\leq \int_{|\lambda|\leq 1} \frac{C_\alpha|\lambda|^\alpha \int_{\mathbb{R}^d} g(\sigma)|^\alpha d\sigma + \int_{\mathbb{R}^d} H_\epsilon(\lambda g(\sigma))d\sigma}{|\lambda|^{1+\beta}}\,d\lambda < +\infty.
$$

Please note that

$$
\sup_{0<\epsilon<1} \int_{|\lambda|\leq 1} \frac{\int_{\mathbb{R}^d} H_\epsilon(\lambda g(\sigma))d\sigma}{|\lambda|^{1+\beta}}\,d\lambda < +\infty
$$

because of (A.3). Hence (109) is proved.

Proof of Lemma 5.5

We first prove the convergence of $\mathbb{E}S_n(y)$. By (2)

$$
\mathbb{E}S_n(y) = \exp\left\{\int_{\mathbb{R}^d\times\mathbb{R}} [\exp(iuy2^{n\tilde{H}}\Delta G_{0,n}(s)) - 1 - iuy2^{n\tilde{H}}\Delta G_{0,n}(s)] \right.
$$
$$
\left. \times\, ds\mu(du)\right\}.
$$

The change of variables $s = \dfrac{\sigma}{2^n}$, $v = uy2^{n\frac{d}{\alpha}}$ leads to

$$
\mathbb{E}S_n(y) = \exp\left\{|y|^\alpha \int_{\mathbb{R}^d\times\mathbb{R}} [\exp(iv\Delta G_{0,1}(\sigma)) - 1 - iv\Delta G_{0,1}(\sigma)] \right.
$$
$$
\left. \times\, d\sigma \frac{dv}{|v|^{1+\alpha}} \mathbf{1}_{|v|\leq|y|2^{n\frac{d}{\alpha}}}\right\}.
$$

and the convergence of $\mathbb{E}S_n(y)$ toward $S(y)$ is proved by using the same arguments as in (61).

Let us now study the variance of $S_n(y)$.

$$
\mathrm{var}S_n(y) = \frac{1}{(2^n - K)^{2d}} \sum_{\mathbf{p},\mathbf{p}'=1}^{2^n-K} I_{\mathbf{p},\mathbf{p}'},
$$

with

$$I_{\mathbf{p},\mathbf{p}'} = \mathbb{E}\exp(iy2^{n\tilde{H}}(\Delta X_{\mathbf{p},n} - \Delta X_{\mathbf{p}',n}))$$
$$- \mathbb{E}\exp(iy2^{n\tilde{H}}\Delta X_{\mathbf{p},n})\,\mathbb{E}\exp(-iy2^{n\tilde{H}}\Delta X_{\mathbf{p}',n}).$$

Because of (2) $I_{\mathbf{p},\mathbf{p}'}$ can also be written

$$I_{\mathbf{p},\mathbf{p}'} = \exp\{\int_{\mathbb{R}^d\times\mathbb{R}}[\exp(iuy2^{n\tilde{H}}(\Delta G_{\mathbf{p},n}(s) - \Delta G_{\mathbf{p}',n}(s))) - 1$$
$$- iuy2^{n\tilde{H}}(\Delta G_{\mathbf{p},n}(s) - \Delta G_{\mathbf{p}',n}(s))]\,ds\mu(du)\}$$
$$- \exp\{\int_{\mathbb{R}^d\times\mathbb{R}}[\exp(iuy2^{n\tilde{H}}\Delta G_{\mathbf{p},n}(s)) + \exp(-iuy2^{n\tilde{H}}\Delta G_{\mathbf{p}',n}(s))$$
$$- 2 - iuy2^{n\tilde{H}}\Delta G_{\mathbf{p},n}(s) - iuy2^{n\tilde{H}}\Delta G_{\mathbf{p}',n}(s)]\,ds\mu(du)\}.$$

Hence $I_{\mathbf{p},\mathbf{p}'} = A_{\mathbf{p},\mathbf{p}'} \times B_{\mathbf{p},\mathbf{p}'}$ with

$$A_{\mathbf{p},\mathbf{p}'} = \exp\left\{\int_{\mathbb{R}^d\times\mathbb{R}}[\exp(iuy2^{n\tilde{H}}\Delta G_{\mathbf{p},n}(s)) + \exp(-iuy2^{n\tilde{H}}\Delta G_{\mathbf{p}',n}(s)) - 2\right.$$
$$\left. -iuy2^{n\tilde{H}}\Delta G_{\mathbf{p},n}(s) - iuy2^{n\tilde{H}}\Delta G_{\mathbf{p}',n}(s)]\,ds\mu(du)\right\}.$$

and

$$B_{\mathbf{p},\mathbf{p}'} = \exp\left\{\int_{\mathbb{R}^d\times\mathbb{R}}[\exp(iuy2^{n\tilde{H}}\Delta G_{\mathbf{p},n}(s)) - 1]\right.$$
$$\left.[\exp(-iuy2^{n\tilde{H}}\Delta G_{\mathbf{p}',n}(s)) - 1]\,ds\mu(du)\right\} - 1.$$

Clearly, one has $|A_{\mathbf{p},\mathbf{p}'}| \leq 1$. The change of variables $s = \dfrac{\sigma}{2^n}$, $v = uy2^{n\frac{d}{\alpha}}$ leads to

$$B_{\mathbf{p},\mathbf{p}'} = \exp\left\{|y|^\alpha\int_{\mathbb{R}^d\times\mathbb{R}}[\exp(iv\Delta G_{\mathbf{p},1}(\sigma)) - 1]\right.$$
$$\left.[\exp(-iv\Delta G_{\mathbf{p}',1}(\sigma)) - 1]d\sigma\frac{dv}{|v|^{1+\alpha}}\mathbf{1}_{|v|\leq 2^n\frac{d}{\alpha}}\right\} - 1.$$

Define:

$$C_{\mathbf{p},\mathbf{p}'} = \int_{\mathbb{R}^d\times\mathbb{R}}[\exp(iv\Delta G_{\mathbf{p}-\mathbf{p}',1}(\sigma)) - 1][\exp(-iv\Delta G_{\mathbf{0},1}(\sigma)) - 1]d\sigma\frac{dv}{|v|^{1+\alpha}},$$

this leads to:

$$B_{\mathbf{p},\mathbf{p}'} = (\exp(C_{\mathbf{p},\mathbf{p}'}|y|^{\alpha}) - 1)(1 + o(1)).$$

We split $C_{\mathbf{p},\mathbf{p}'}$ into two parts, with A to be chosen later:

$$T_1 = \int_{\mathbb{R}^d \times \{|v| \leq A\}} [\exp(iv \Delta G_{\mathbf{p}-\mathbf{p}',1}(\sigma)) - 1][\exp(-iv \Delta G_{\mathbf{0},1}(\sigma)) - 1] d\sigma \frac{dv}{|v|^{1+\alpha}},$$

$$T_2 = \int_{\mathbb{R}^d \times \{|v| \geq A\}} [\exp(iv \Delta G_{\mathbf{p}-\mathbf{p}',1}(\sigma)) - 1][\exp(-iv \Delta G_{\mathbf{0},1}(\sigma)) - 1] d\sigma \frac{dv}{|v|^{1+\alpha}},$$

$$|T_1| \leq \int_{\mathbb{R}^d \times |v| \leq A} \left| \sum_{\ell=0}^{K} a_\ell \|\ell + \mathbf{p} - \mathbf{p}' - \sigma\|^{H-\frac{d}{2}} \sum_{\ell=0}^{K} a_\ell \|\ell - \sigma\|^{H-\frac{d}{2}} \right| d\sigma \frac{dv}{|v|^{\alpha-1}}$$

$$\leq CA^{2-\alpha} \int_{\mathbb{R}^d} \left| \sum_{\ell=0}^{K} a_\ell \|\ell + \mathbf{p} - \mathbf{p}' - \sigma\|^{H-\frac{d}{2}} \sum_{\ell=0}^{K} a_\ell \|\ell - \sigma\|^{H-\frac{d}{2}} \right| d\sigma.$$

A Taylor expansion of order 2 around $\mathbf{p} - \mathbf{p}'$ of

$$x \mapsto \int_{\mathbb{R}^d} \left| \sum_{\ell=0}^{K} a_\ell \|\ell + x - \sigma\|^{H-\frac{d}{2}} \sum_{\ell=0}^{K} a_\ell \|\ell - \sigma\|^{H-\frac{d}{2}} \right| d\sigma$$

is used. Hence

$$|T_1| \leq CA^{2-\alpha} \|\mathbf{p} - \mathbf{p}'\|^{H-\frac{d}{2}-2}.$$

Moreover, for $\delta > 0$ arbitrarily small:

$$|T_2| \leq \frac{C}{A^{\alpha-\delta}} \int_{\mathbb{R}^d \times \mathbb{R}} |[\exp(iv \Delta G_{\mathbf{p}-\mathbf{p}',1}(\sigma)) - 1][\exp(-iv \Delta G_{\mathbf{0},1}(\sigma)) - 1]|$$

$$\times d\sigma \frac{dv}{|v|^{1+\delta}}.$$

By Cauchy–Schwarz,

$$\left(\int_{\mathbb{R}^d} |[\exp(iv \Delta G_{\mathbf{p}-\mathbf{p}',1}(\sigma)) - 1][\exp(-iv \Delta G_{\mathbf{0},1}(\sigma)) - 1]| d\sigma \right)^2 \leq$$

$$\int_{\mathbb{R}^d} |[\exp(iv \Delta G_{\mathbf{p}-\mathbf{p}',1}(\sigma)) - 1]|^2 d\sigma \int_{\mathbb{R}^d} |[\exp(-iv \Delta G_{\mathbf{0},1}(\sigma)) - 1]|^2 d\sigma$$

$$= \left(\int_{\mathbb{R}^d} |[\exp(-iv \Delta G_{\mathbf{0},1}(\sigma)) - 1]|^2 d\sigma \right)^2,$$

so that

$$|T_2| \leq \frac{C'}{A^{\alpha-\delta}}.$$

We choose A such that $A^{2+\delta} = ||\mathbf{p} - \mathbf{p}'||^{2+\frac{d}{2}-H}$. Therefore, as $||\mathbf{p} - \mathbf{p}'|| \to +\infty$:

$$|B_{\mathbf{p},\mathbf{p}'}| \leq C||\mathbf{p} - \mathbf{p}'||^{(\delta-\alpha)\frac{2+\frac{d}{2}-H}{2+\delta}}|y|^{\alpha}.$$

We choose δ small enough, so that $\sum_n \mathrm{var} S_n(y)$ is convergent and Borel–Cantelli's Lemma concludes the proof of the Lemma 5.5.

Proof of Theorem 6.1

Since $0 \leq r \wedge 2 \leq r$ and $\mathbb{E}(|X_1|^r) < +\infty$, we also have that $\mathbb{E}\left(|X_1|^{r\wedge 2}\right) < +\infty$. Then, we can assume that $r \leq 2$.

Set $R_N = \sum_{n=N+1}^{+\infty} Y_n X_n$ and $r = r \wedge 2 \in (0,2)$. Then, let us fix $M > 0$ and set

$$\Omega_M = \left\{\sup_{n \geq 1} n^{-1/r}|X_n| \leq M\right\}.$$

Hence for any $\varepsilon > 0$,

$$\mathbb{P}\left(\Omega_M \cap \{|R_N| \geq N^{-\varepsilon}\}\right) \leq \mathbb{P}\left(\left|\sum_{n=N+1}^{+\infty} Y_n W_n\right| \geq N^{-\varepsilon}\right),$$

where $W_n = X_n \mathbf{1}_{|X_n| \leq M n^{1/r}}$. Since X_n, $n \geq 1$, are i.i.d. and symmetric, $(W_n)_{n \geq 1}$ is a sequence of independent symmetric random variables. Then, since $(Y_n)_{n \geq 1}$ satisfies the assumption (115) and is independent of $(W_n)_{n \geq 1}$, by the contraction principle for symmetric random variables sequences, see [23] page 95,

$$\mathbb{P}\left(\Omega_M \cap \{|R_N| \geq N^{-\varepsilon}\}\right) \leq 2\mathbb{P}\left(C\left|\sum_{n=N+1}^{+\infty} T_n^{-1/\gamma} W_n\right| \geq N^{-\varepsilon}\right).$$

Hence,

$$\mathbb{P}\left(\Omega_M \cap \{|R_N| \geq N^{-\varepsilon}\}\right) \leq 2\mathbb{P}\left(\sup_{n \geq N+1} \frac{n}{T_n} \geq 10\right) + 2A_N, \qquad (A.8)$$

where

$$A_N = \mathbb{P}\left(\left\{\sup_{n \geq N+1} \frac{n}{T_n} < 10\right\} \cap \left\{C \left|\sum_{n=N+1}^{+\infty} T_n^{-1/\gamma} W_n\right| \geq N^{-\varepsilon}\right\}\right).$$

Step 1

$$\mathbb{P}\left(\sup_{n \geq N+1} \frac{n}{T_n} \geq 10\right) \leq \sum_{n=N+1}^{+\infty} \mathbb{P}(T_n \leq n/10) \leq \sum_{n=N+1}^{+\infty} \frac{n^n}{10^n n!}.$$

Hence, by the Stirling formula,

$$\mathbb{P}\left(\sup_{n \geq N+1} \frac{n}{T_n} \geq 10\right) \leq C_1 \exp\left(-C_2 N\right), \tag{A.9}$$

with $C_1 > 0$ and $C_2 > 0$.

Step 2 By the assumptions of independence, $(T_n)_{n \geq 1}$ and $(W_n)_{n \geq 1}$ are independent. Therefore, by the contraction principle for symmetric random variables sequences,

$$A_N \leq 2\mathbb{P}\left(C \left|\sum_{n=N+1}^{+\infty} n^{-1/\gamma} W_n\right| \geq 10^{-1/\gamma} N^{-\varepsilon}\right).$$

Furthermore, by independence and symmetry,

$$A_N \leq 4\mathbb{P}\left(C \sum_{n=N+1}^{+\infty} n^{-1/\gamma} W_n \geq 10^{-1/\gamma} N^{-\varepsilon}\right)$$

$$\leq 4\exp\left(-\frac{10^{-1/\gamma} \lambda N^{-\varepsilon}}{C}\right) \prod_{n=N+1}^{+\infty} \mathbb{E}\left(\exp\left(\lambda n^{-1/\gamma} W_n\right)\right), \tag{A.10}$$

since $\mathbb{P}(\xi \geq x) \leq e^{\lambda x} \mathbb{E}(e^{\lambda \xi})$, for all $\lambda > 0$. Moreover, since W_n is a symmetric random variable,

$$\mathbb{E}\left(\exp\left(\lambda n^{-1/\gamma} W_n\right)\right) = 1 + \sum_{j=1}^{+\infty} \frac{\lambda^{2j}}{(2j)!} n^{-2j/\gamma} \mathbb{E}(W_n^{2j}).$$

Then let $a = 1/\gamma - 1/r$ and $n \geq N + 1$. Note that for $j \geq 1$, $2j \geq r$ and

$$\mathbb{E}(W_n^{2j}) \leq \mathbb{E}(|X_1|^r)\left(M n^{1/r}\right)^{2j-r}.$$

Therefore,

$$\mathbb{E}\left(\exp\left(\lambda n^{-1/\gamma}W_n\right)\right) \leq 1 + \frac{\mathbb{E}(|X_1|^r)\lambda^2 M^{2-r}\exp\left(\lambda^2 M^2 n^{-2a}\right)}{2n^{1+2a}}$$

$$\leq \exp\left(\frac{\mathbb{E}(|X_1|^r)\lambda^2 M^{2-r}\exp\left(\lambda^2 M^2 N^{-2a}\right)}{2n^{1+2a}}\right).$$

As a consequence, taking $\lambda = 10^{1/\gamma}N^a$ in (A.10), there exist $C_3 > 0$ and $C_4 > 0$, which do not depend on N, such that

$$A_N \leq C_3 \exp\left(-C_4 N^{a-\varepsilon}\right). \tag{A.11}$$

Step 3 In view of (A.8), (A.9) and (A.11), for every $M > 0$ and every $\varepsilon \in (0, 1/\gamma - 1/r)$,

$$\sum_{N=1}^{+\infty} \mathbb{P}\left(\Omega_M \cap \{|R_N| \geq N^{-\varepsilon}\}\right) < +\infty.$$

Hence, by the Borel Cantelli lemma, for almost all $\omega \in \Omega_M$,

$$\sup_{N\geq 1} N^\varepsilon |R_N| < +\infty.$$

Furthermore, we have

$$\mathbb{P}(\Omega_M^c) \leq \sum_{n=1}^{+\infty} \mathbb{P}\left(|X_n| > Mn^{1/r}\right) \leq \frac{1}{M^r}\mathbb{E}(|X_1|^r) \to 0$$

as $M \to +\infty$ and thus $\lim_{M\to+\infty} \mathbb{P}(\Omega_M) = 1$. Then, for every $\varepsilon \in (0, 1/\gamma - 1/r)$, almost surely,

$$\sup_{N\geq 1} N^\varepsilon |R_N| < +\infty,$$

which concludes the proof.

Proof of Theorem 6.2

It is a simple modification of the proof of Theorem 6.1.

Let $M > 0$, $\Omega_M = \left\{\sup_{n\geq 1}\left|n^{-1/r}X_n\right| \leq M\right\}$, $W_n = X_n\mathbf{1}_{|X_n|\leq n^{1/r}M}$ and

$$R_N = \sum_{n=N+1}^{+\infty} T_n^{-1/\gamma}|X_n|.$$

As in proof of Theorem 6.1,

$$\mathbb{P}\big(\Omega_M \cap \{|R_N| \geq N^{-\varepsilon}\}\big) \leq \mathbb{P}\bigg(\sup_{n \geq N+1} \frac{n}{T_n} \geq 10\bigg) + A_N \qquad (A.12)$$

where

$$A_N = \mathbb{P}\bigg(\bigg\{\sup_{n \geq N+1} \frac{n}{T_n} < 10\bigg\} \cap \bigg\{\sum_{n=N+1}^{+\infty} T_n^{-1/\gamma}|W_n| \geq N^{-\varepsilon}\bigg\}\bigg).$$

Remark now that the contraction principle used in the proof of Theorem 6.1 can not be applied since $|W_n|$ is not a symmetric random variable. However, since $|W_n| \geq 0$,

$$A_N \leq \mathbb{P}\bigg(\sum_{n=N+1}^{+\infty} n^{-1/\gamma}|W_n| \geq 10^{-1/\gamma}N^{-\varepsilon}\bigg)$$
$$\leq \exp\big(-10^{-1/\gamma}\lambda N^{-\varepsilon}\big) \prod_{n=N+1}^{+\infty} \mathbb{E}\big(\exp\big(\lambda n^{-1/\gamma}|W_n|\big)\big),$$

where $\lambda > 0$. Furthermore,

$$\mathbb{E}\big(\exp\big(\lambda n^{-1/\gamma}|W_n|\big)\big) = 1 + \sum_{j=1}^{+\infty} \frac{\lambda^j}{j!} n^{-j/\gamma} \mathbb{E}\big(|W_n|^j\big)$$
$$\leq 1 + \mathbb{E}(|X_1|^r) \sum_{j=1}^{+\infty} \frac{\lambda^j}{j!} n^{-j/\gamma}\big(Mn^{1/r}\big)^{j-r} \quad \text{since } r \leq 1$$
$$\leq \exp\big(\mathbb{E}(|X_1|^r)\lambda n^{-1-a}M^{1-r}\exp\big(\lambda MN^{-a}\big)\big),$$

where $a = 1/\gamma - 1/r$ and $n \geq N + 1$. Hence, choosing $\lambda = 10^{1/\gamma}N^a$, there exists C, which does not depend on N, such that

$$A_N \leq C \exp\big(-N^{a-\varepsilon}\big).$$

Consequently, the arguments used in step 3 of the proof of Theorem 6.1 lead to the conclusion.

Acknowledgements The author would like to thank Claudia Küppelberg for the friendly pressure she put on him. Without this help and her patience he is doubtful that this survey would have been ever written. I would also like to thank both referees for their careful reading that improved my first version.

References

1. P. Abry, P. Gonçalves, J. Lévy Véhel (eds.), in *Scaling, Fractals and Wavelets*. Digital Signal and Image Processing Series (ISTE, London, 2009)
2. S. Asmussen, J. Rosiński, Approximations of small jumps of Lévy processes with a view towards simulation. J. Appl. Probab. **38**(2), 482–493 (2001)
3. A. Ayache, A. Benassi, S. Cohen, J.L. Véhel, in *Regularity and Identification of Generalized Multifractional Gaussian Processes*. Séminaire de Probabilité XXXVIII. Lecture Notes in Mathematics, vol. 1804 (Springer Verlag, 2004)
4. J.M. Bardet, Testing for the presence of self-similarity of gaussian time series having stationary increments. J. Time Ser. Anal. **21**(5), 497–515 (2000)
5. J.-M. Bardet, G. Lang, G. Oppenheim, A. Philippe, M.S. Taqqu, in *Generators of Long-Range Dependent Processes: A Survey*. Theory and Applications of Long-Range Dependence (Birkhäuser, Boston, 2003), pp. 579–623
6. A. Benassi, S. Jaffard, D. Roux, Gaussian processes and pseudodifferential elliptic operators. Revista Mathematica Iberoamericana **13**(1), 19–90 (1996)
7. A. Benassi, S. Cohen, J. Istas, Identification and properties of real harmonizable fractional Lévy motions. Bernoulli **8**(1), 97–115 (2002)
8. J.-F. Coeurjolly, Estimating the parameters of a fractional Brownian motion by discrete variations of its sample paths. Stat. Inference Stochast. Process. **4**(2), 199–227 (2001)
9. S. Cohen, J. Istas, Fractional fields and applications (2011), http://www.math.univ-toulouse.fr/~cohen
10. S. Cohen, M.S. Taqqu, Small and large scale behavior of the Poissonized telecom process. Meth. Comput. Appl. Probab. **6**(4), 363–379 (2004)
11. L. Debnath (ed.), in *Wavelets and Signal Processing*. Applied and Numerical Harmonic Analysis (Birkhäuser, Boston, 2003)
12. M. Dekking, J.L. Véhel, E. Lutton, C. Tricot (eds.), *Fractals: Theory and Applications in Engineering* (Springer, Berlin, 1999)
13. M.E. Dury, Identification et simulation d'une classe de processus stables autosimilaires à accroissements stationnaires. Ph.D. thesis, Université Blaise Pascal, Clermont-Ferrand (2001)
14. P. Flandrin, P. Abry, in *Wavelets for Scaling Processes*. Fractals: Theory and Applications in Engineering (Springer, London, 1999), pp. 47–64
15. J. Istas, G. Lang, Quadratic variations and estimation of the Holder index of a gaussian process. Ann. Inst. Poincaré **33**(4), 407–436 (1997)
16. A. Janssen, Zero-one laws for infinitely divisible probability measures on groups. Zeitschrift für Wahrscheinlichkeitstheorie **60**, 119–138 (1982)
17. I. Kaj, M. Taqqu, in *Convergence to Fractional Brownian Motion and to the Telecom Process: The Integral Representation Approach*, ed. by M.E. Vares, V. Sidoravicius. Brazilian Probability School, 10th anniversary volume (Birkhauser, Boston, 2007)
18. O. Kallenberg, in *Foundations of Modern Probability*, 2nd edn. Probability and Its Applications (New York) (Springer, New York, 2002)
19. A. Kolmogorov, Wienersche Spiralen und einige andere interessante Kurven in Hilbertsche Raum. C. R. (Dokl.) Acad. Sci. URSS **26**, 115–118 (1940)
20. H. Kunita, in *Stochastic Flows and Stochastic Differential Equations*, vol. 24 of Cambridge Studies in Advanced Mathematics (Cambridge University Press, Cambridge, 1997). Reprint of the 1990 original
21. C. Lacaux, Real harmonizable multifractional Lévy motions. Ann. Inst. H. Poincaré Probab. Statist. **40**(3), 259–277 (2004)
22. C. Lacaux, Series representation and simulation of multifractional Lévy motions. Adv. Appl. Probab. **36**(1), 171–197 (2004)
23. M. Ledoux, M. Talagrand, in *Probability in Banach Spaces*, vol. 23 of Ergebnisse der Mathematik und ihrer Grenzgebiete (3) [Results in Mathematics and Related Areas (3)]. Isoperimetry and Processes (Springer, Berlin, 1991)

24. S. Mallat, A theory of multiresolution signal decomposition: The wavelet representation. IEEE Trans. Pattern Anal. Mach. Intell. **11**, 674–693 (1989)
25. B. Mandelbrot, J. Van Ness, Fractional Brownian motion, fractional noises and applications. Siam Rev. **10**, 422–437 (1968)
26. T. Marquardt, Fractional Lévy processes with an application to long memory moving average processes. Bernoulli **12**(6), 1099–1126 (2006)
27. R. Peltier, J. Lévy-Vehel, Multifractional Brownian motion: Definition and preliminary results (1996). Prepublication, http://www-syntim.inria.fr/fractales/
28. V.V. Petrov, An estimate of the deviation of the distribution of independent random variables from the normal law. Soviet Math. Doklady **6**, 242–244 (1982)
29. P. Protter, *Stochastic Integration and Differential Equations* (Springer, Berlin, 1990)
30. B.S. Rajput, J. Rosiński, Spectral representations of infinitely divisible processes. Probab. Theor. Relat. Fields **82**(3), 451–487 (1989)
31. J. Rosinski, On path properties of certain infinitely divisible process. Stochast. Process. Appl. **33**, 73–87 (1989)
32. J. Rosiński, On series representations of infinitely divisible random vectors. Ann. Probab. **18**(1), 405–430 (1990)
33. J. Rosiński, in *Series Representations of Lévy Processes from the Perspective of Point Processes*. Lévy Processes (Birkhäuser, Boston, 2001), pp. 401–415
34. G. Samorodnitsky, M.S. Taqqu, *Stable Non-Gaussian Random Processes* (Chapman and Hall, London, 1994)
35. K.-I. Sato, in *Lévy Processes and Infinitely Divisible Distributions*, vol. 68 of Cambridge Studies in Advanced Mathematics (Cambridge University Press, Cambridge, 1999). Translated from the 1990 Japanese original, Revised by the author
36. E. Simoncelli, in *Bayesian Denoising of Visual Images in the Wavelet Domain*. Lect. Notes Stat., vol. 141 (Springer, Berlin, 1999), pp. 291–308
37. S. Stoev, M.S. Taqqu, Simulation methods for linear fractional stable motion and FARIMA using the fast Fourier transform. Fractals **12**(1), 95–121 (2004)
38. S. Stoev, M.S. Taqqu, Stochastic properties of the linear multifractional stable motion. Adv. Appl. Probab. **36**(4), 1085–1115 (2004)
39. S. Stoev, M.S. Taqqu, Path properties of the linear multifractional stable motion. Fractals **13**(2), 157–178 (2005)
40. S. Stoev, M.S. Taqqu, How rich is the class of multifractional Brownian motion. Stochast. Process. Appl. **116**(1), 200–221 (2006)
41. B. Vidakovic, *Statistical Modeling by Wavelets* (Wiley, New York, 1999)
42. W.B. Wu, G. Michailidis, D. Zhang, Simulating sample paths of linear fractional stable motion. IEEE Trans. Inform. Theor. **50**(6), 1086–1096 (2004)

The Theory of Scale Functions for Spectrally Negative Lévy Processes

Alexey Kuznetsov, Andreas E. Kyprianou and Victor Rivero

Abstract The purpose of this review article is to give an up to date account of the theory and applications of scale functions for spectrally negative Lévy processes. Our review also includes the first extensive overview of how to work numerically with scale functions. Aside from being well acquainted with the general theory of probability, the reader is assumed to have some elementary knowledge of Lévy processes, in particular a reasonable understanding of the Lévy–Khintchine formula and its relationship to the Lévy–Itô decomposition. We shall also touch on more general topics such as excursion theory and semi-martingale calculus. However, wherever possible, we shall try to focus on key ideas taking a selective stance on the technical details. For the reader who is less familiar with some of the mathematical theories and techniques which are used at various points in this review, we note that all the necessary technical background can be found in the following texts on Lévy processes; (Bertoin, Lévy Processes (1996); Sato, Lévy Processes and Infinitely Divisible Distributions (1999); Kyprianou, Introductory Lectures on Fluctuations of Lévy Processes and Their Applications (2006); Doney, Fluctuation Theory for Lévy Processes (2007)), Applebaum Lévy Processes and Stochastic Calculus (2009).

Mathematics Subject Classification 2000: Primary: 60G40, 60G51, 60J45
Secondary: 91B70

A. Kuznetsov
Department of Mathematics and Statistics, York University, 4700 Keele Street, Toronto, ON, Canada M3J 1P3
e-mail: kuznetsov@mathstat.yorku.ca; http://www.math.yorku.ca/~akuznets/

A.E. Kyprianou (✉)
Department of Mathematical Sciences, University of Bath, Claverton Down, Bath BA1 2UU, UK
e-mail: a.kyprianou@bath.ac.uk; http://www.maths.bath.ac.uk/~ak257/

V. Rivero
Centro de Investigación en Matemáticas A.C., Calle Jalisco s/n, C.P. 36240 Guanajuato, Gto.
Mexico
e-mail: rivero@cimat.mx; http://www.cimat.mx/~rivero/

Keywords Applied probability • Excursion theory • First passage problem • Fluctuation theory • Laplace transform • Scale functions • Spectrally negative Lévy processes

1 Motivation

1.1 Spectrally Negative Lévy Processes

Let us begin with a brief overview of what is to be understood by a spectrally negative Lévy process. Suppose that $(\Omega, \mathscr{F}, \mathbb{F}, P)$ is a filtered probability space with filtration $\mathbb{F} = \{\mathscr{F}_t : t \geq 0\}$ which is naturally enlarged (cf. Definition 1.3.38 of [18]). A Lévy process on this space is a strong Markov, $\mathbb{F}$-adapted process $X = \{X_t : t \geq 0\}$ with càdlàg paths having the properties that $P(X_0 = 0) = 1$ and for each $0 \leq s \leq t$, the increment $X_t - X_s$ is independent of $\mathscr{F}_s$ and has the same distribution as X_{t-s}. In this sense, it is said that a Lévy process has stationary independent increments.

On account of the fact that the process has stationary and independent increments, it is not too difficult to show that there is a function $\Psi : \mathbb{R} \mapsto \mathbb{C}$, such that

$$E\left(e^{i\theta X_t}\right) = e^{-t\Psi(\theta)}, \qquad t \geq 0, \theta \in \mathbb{R},$$

where E denotes expectation with respect to P. The Lévy–Khintchine formula gives the general form of the function Ψ. That is,

$$\Psi(\theta) = i\mu\theta + \frac{\sigma^2}{2}\theta^2 + \int_{(-\infty,\infty)} \left(1 - e^{i\theta x} + i\theta x \, \mathbf{1}_{|x|<1}\right) \Pi(dx), \qquad (1)$$

for every $\theta \in \mathbb{R}$, where $\mu \in \mathbb{R}$, $\sigma \in \mathbb{R}$ and Π is a measure on $\mathbb{R} \backslash \{0\}$ such that $\int (1 \wedge x^2) \Pi(dx) < \infty$. Often we wish to specify the law of X when issued from $x \in \mathbb{R}$. In that case we write P_x, still reserving the notation P for the special case P_0. Note in particular that X under P_x has the same law as $x + X$ under P. The notation E_x will be used for the obvious expectation operator.

We say that X is *spectrally negative* if the measure Π is carried by $(-\infty, 0)$, that is $\Pi(0, \infty) = 0$. We exclude from the discussion however the case of monotone paths. Under the present circumstances that means we are excluding the case of a descending subordinator and the case of a pure positive linear drift. Included in the discussion however are the difference of these two processes, as well as a Brownian motion with drift. Also included are processes such as asymmetric α-stable processes for $\alpha \in (1, 2)$ which have unbounded variation and zero quadratic variation. These processes are by no means representative of the true variety of processes that populate the class of spectrally negative Lévy processes.

Indeed in the forthcoming text we shall see many different explicit examples of spectrally negative Lévy processes. Moreover, by adding independent copies of any of the aforementioned, and/or other, spectrally negative processes together, the resulting process remains within the class of spectrally negative Lévy process. Notationally we say that a Lévy process X is *spectrally positive* when $-X$ is spectrally negative.

Thanks to the fact that there are no positive jumps, it is possible to talk of the Laplace exponent $\psi(\lambda)$ for a spectrally negative Lévy process, defined by

$$E\left(e^{\lambda X_t}\right) = e^{\psi(\lambda)t}, \tag{2}$$

for at least $\lambda \geq 0$. In other words, taking into account a straightforward analytical extension of the characteristic exponent, we have $\psi(\lambda) = -\Psi(-i\lambda)$. In particular

$$\psi(\lambda) = -\mu\lambda + \frac{\sigma^2}{2}\lambda^2 + \int_{(-\infty,0)} \left(e^{\lambda x} - 1 - \lambda x\, \mathbf{1}_{|x|<1}\right) \Pi(dx). \tag{3}$$

Using Hölder's inequality, or alternatively differentiating, it is easy to check that ψ is strictly convex and tends to infinity as λ tends to infinity. This allows us to define for $q \geq 0$,

$$\Phi(q) = \sup\left\{\lambda \geq 0 : \psi(\lambda) = q\right\},$$

the largest root of the equation $\psi(\lambda) = q$. Note that there can exist two roots when $q = 0$ (there is always a root at zero on account of the fact that $\psi(0) = 0$) and precisely one root when $q > 0$. Further, from a straightforward differentiation of (2), we can identify $\psi'(0^+) = E(X_1) \in [-\infty, \infty)$ which determines the long term behaviour of the process. Indeed when $\pm\psi'(0+) > 0$ we have $\lim_{t\uparrow\infty} X_t = \pm\infty$, that is to say that the process drifts towards $\pm\infty$, and when $\psi'(0+) = 0$ then $\limsup_{t\uparrow\infty} X_t = -\liminf_{t\uparrow\infty} X_t = \infty$, in other words the process oscillates.

Further details about the class of spectrally negative Lévy processes and how they embed within the general class of Lévy processes can be found in the monographs of Bertoin [15], Sato [91], Kyprianou [66] and Applebaum [7].

1.2 Scale Functions and Applied Probability

Let us now turn our attention to the definition of *scale functions* and motivate the need for a theory thereof.

Definition 1.1. For a given spectrally negative Lévy process, X, with Laplace exponent ψ, we define a family of functions indexed by $q \geq 0$, $W^{(q)} : \mathbb{R} \to [0, \infty)$, as follows. For each given $q \geq 0$, we have $W^{(q)}(x) = 0$ when $x < 0$, and otherwise on $[0, \infty)$, $W^{(q)}$ is the unique right continuous function whose Laplace transform is given by

$$\int_0^\infty e^{-\beta x} W^{(q)}(x) \mathrm{d}x = \frac{1}{\psi(\beta) - q} \tag{4}$$

for $\beta > \Phi(q)$. For convenience we shall always write W in place of $W^{(0)}$. Typically we shall refer the functions $W^{(q)}$ as *q-scale functions*, but we shall also refer to W as just the *scale function*.

From the above definition alone, it is not clear why one would be interested in such functions. Later on we shall see that scale functions appear in the vast majority of known identities concerning boundary crossing problems and related path decompositions. This in turn has consequences for their use in a number of classical applied probability models which rely heavily on such identities. To give but one immediate example of a boundary crossing identity which necessitates Definition 1.1, consider the following result, the so-called *two-sided exit problem*, which has a long history; see for example [16, 17, 88, 100, 107].

Theorem 1.2. *Define*

$$\tau_a^+ = \inf\{t > 0 : X_t > a\} \text{ and } \tau_0^- = \inf\{t > 0 : X_t < 0\}.$$

For all $q \geq 0$, $a > 0$ and $x < a$,

$$E_x\left(e^{-q\tau_a^+} 1_{\{\tau_a^+ < \tau_0^-\}}\right) = \frac{W^{(q)}(x)}{W^{(q)}(a)}. \tag{5}$$

In fact, it is through this identity that the "scale function" gets its name. Possibly the first reference to this terminology is found in Bertoin [14]. Indeed (5) has some mathematical similarities with an analogous identity which holds for a large class of one-dimensional diffusions and which involves so-called scale functions for diffusions; see for example Proposition VII, 3.2 of Revuz and Yor [86]. In older Soviet-Ukranian literature $W^{(q)}$ is simply referred to as a *resolvent*, see for example [56].

Example 1.3. Let us consider the case when X is a Brownian motion with drift and compound Poisson jumps

$$X_t = \sigma B_t + \mu t - \sum_{i=1}^{N_t} \xi_i, \quad t \geq 0,$$

where ξ_i are i.i.d. random variables which are exponentially distributed with parameter $\rho > 0$ and N_t is an independent Poisson process with intensity $a > 0$. We see that the Lévy measure of X is finite, thus we can replace the cutoff function $h(x) = 1_{|x|<1}$ by $h(x) \equiv 0$ in the Lévy–Khintchine formula (3), which changes the value of μ but nonetheless gives us a simple expression for the Laplace exponent of X;

$$\psi(z) = \frac{\sigma^2}{2}z^2 + \mu z - \frac{az}{\rho + z}, \qquad z \geq 0. \tag{6}$$

By considering the behavior of $\psi(z)$ as $z \to \pm\infty$ and $z \to \rho^\pm$ it is easy to verify that for every $q > 0$ the equation $\psi(z) = q$ has exactly three real solutions $\{-\zeta_2, -\zeta_1, \Phi(q)\}$, which satisfy

$$-\zeta_2 < -\rho < -\zeta_1 < 0 < \Phi(q).$$

Note that the equation $\psi(z) = q$ can be rewritten as a cubic equation, and the solutions to this equation can be found explicitly in terms of coefficients $(\sigma, \mu, a, \rho, q)$ by using Cardano's formulas. However, even without writing down explicit expressions for ζ_i and $\Phi(q)$ we can still obtain a convenient expression for the scale function by considering the partial fraction decomposition of the rational function $1/(\psi(z) - q)$ and inverting the Laplace transform in (4). We conclude that

$$W^{(q)}(x) = \frac{e^{\Phi(q)x}}{\psi'(\Phi(q))} + \frac{e^{-\zeta_1 x}}{\psi'(-\zeta_1)} + \frac{e^{-\zeta_2 x}}{\psi'(-\zeta_2)}, \qquad x \geq 0. \tag{7}$$

Let us consider two special cases, when the expression (7) can be simplified. For the first example, we assume that X has no Gaussian component ($\sigma = 0$). In that case, to avoid the cases that X has monotone paths, we need to further assume that $\mu > 0$. We see that as $\sigma^2 \to 0^+$ we have $\zeta_2 \to +\infty$ and the corresponding term in (7) disappears. At the same time, the equation $\psi(z) = q$ reduces to a *quadratic* equation, and we can calculate explicitly

$$\zeta_1 = \frac{1}{2\mu}\left(\sqrt{(a + q - \mu\rho)^2 + 4\mu q\rho} - (a + q - \mu\rho)\right),$$

$$\Phi(q) = \frac{1}{2\mu}\left(\sqrt{(a + q - \mu\rho)^2 + 4\mu q\rho} + (a + q - \mu\rho)\right).$$

These formulas combined with (7) and (6) provide an explicit formula for the scale function $W^{(q)}(x)$. If we set $q = 0$ we obtain a particularly simple form of the scale function

$$W(x) = \frac{1}{\frac{a}{\rho} - \mu}\left[\frac{a}{\mu\rho}e^{\left(\frac{a}{\rho} - \mu\right)\frac{\rho}{\mu}x} - 1\right].$$

As a second example, we consider the case when X has no jump part, that is X is a Brownian motion with drift. Again, one can see that as $\rho \to +\infty$ we have $\zeta_2 \to +\infty$, so the corresponding term in (7) disappears and the equation

$\psi(z) = q$ reduces to a quadratic equation. The roots of this quadratic equation can be computed explicitly as

$$\zeta_1 = \frac{1}{\sigma^2}\left(\sqrt{\mu^2 + 2q\sigma^2} + \mu\right),$$

$$\Phi(q) = \frac{1}{\sigma^2}\left(\sqrt{\mu^2 + 2q\sigma^2} - \mu\right),$$

and we find that the q-scale function, for $q \geq 0$, is now given by

$$W^{(q)}(x) = \frac{1}{\sqrt{\mu^2 + 2q\sigma^2}}\left[e^{\left(\sqrt{\mu^2+2q\sigma^2}-\mu\right)\frac{x}{\sigma^2}} - e^{-\left(\sqrt{\mu^2+2q\sigma^2}+\mu\right)\frac{x}{\sigma^2}}\right].$$

In the forthcoming chapters we shall explore in more detail the formal analytical properties of scale functions as well as look at explicit examples and methodologies for working numerically with scale functions. However let us first conclude this motivational chapter by giving some definitive examples of how scale functions make a non-trivial contribution to a variety of applied probability models.

Optimal stopping: Suppose that $x, q > 0$ and consider the optimal stopping problem

$$v(x) = \sup_{\tau} E\left(e^{-q\tau + (\overline{X}_\tau \vee x)}\right),$$

where the supremum is taken over all, almost surely finite stopping times with respect to X and $\overline{X}_t = \sup_{s \leq t} X_s$. The solution to this problem entails finding an expression for the value function $v(x)$ and, if the supremum is attained by some stopping time τ^*, describing this optimal stopping time quantitatively.

This particular optimal stopping problem was conceived in connection with the pricing and hedging of certain exotic financial derivatives known as *Russian options* by Shepp and Shiryaev [93,94]. Originally formulated in the Black-Scholes setting, i.e. the case that X is a linear Brownian motion, it is natural to look at the solution of this problem in the case that X is replaced by a Lévy process. This is of particular pertinence on account of the fact that modern theories of mathematical finance accommodate for, and even prefer, such a setting. As a first step in this direction, Avram et al. [9] considered the case that X is a spectrally negative Lévy process. These authors found that scale functions played a very natural role in describing the solution (v, τ^*). Indeed, it was shown under relatively mild additional conditions that

$$v(x) = e^x \left(1 + q \int_0^{x^* - x} W^{(q)}(y)\mathrm{d}y\right)$$

where, thanks to properties of scale functions, it is known that the constant $x^* \in [0, \infty)$ necessarily satisfies

$$x^* = \inf\left\{x \geq 0 : 1 + q \int_0^x W^{(q)}(y)\mathrm{d}y - qW^{(q)}(x) \leq 0\right\}.$$

In particular an optimal stopping time can be taken as

$$\tau^* = \inf\left\{t > 0 : (x \vee \overline{X}_t) - X_t > x^*\right\},$$

which agrees with the original findings of Shepp and Shiryaev [93, 94] in the diffusive case.

Ruin Theory: One arm of actuarial mathematics concerns itself with modelling the event of ruin of an insurance firm. The surplus wealth of the insurance firm is often modelled as the difference of a linear growth, representing the deterministic collection of premiums, and a compound Poisson subordinator, representing a random sequence of i.i.d. claims. It is moreover quite often assumed that the claim sizes are exponentially distributed. Referred to as Cramér–Lundberg processes, such models for the surplus belong to the class of spectrally negative Lévy processes.

Classical ruin theory concerns the distribution of the, obviously named, *time to ruin*,

$$\tau_0^- := \inf\left\{t > 0 : X_t < 0\right\}, \tag{8}$$

where X is the Cramér–Lundberg process. Recent work has focused on more complex additional distributional features of ruin such as the overshoot and undershoot of the surplus at ruin. These quantities correspond to deficit at ruin and the wealth prior to ruin respectively. A large body of literature exists in this direction with many contributions citing as a point of reference the paper Gerber and Shiu [45]. For this reason, the study of the joint distribution of the ruin time as well as the overshoot and undershoot at ruin is often referred to as Gerber–Shiu theory.

It turns out that scale functions again provide a natural tool with which one may give a complete characterisation of the ruin problem in the spirit of Gerber–Shiu theory. Moreover, this can be done for a general spectrally negative Lévy process rather than just the special case of a Cramér–Lundberg process. Having said that, scale functions already serve their purpose in the classical Cramér–Lundberg setting. A good case in point is the following result (cf. [19, 20]) which even goes a little further than what classical Gerber–Shiu theory demands in that it also incorporates distributional information on the infimum of the surplus prior to ruin.

Theorem 1.4. *Suppose that X is a spectrally negative Lévy processes. For $t \geq 0$ define $\underline{X}_t = \inf_{s \leq t} X_s$. Let $q, x \geq 0$, $u, v, y \geq 0$, then*

$$E_x\left(e^{-q\tau_0^-}; -X_{\tau_0^-} \in du, X_{\tau_0^- -} \in dv, \underline{X}_{\tau_0^- -} \in dy\right)$$

$$= 1_{\{0 < y < v \wedge x, u > 0\}} e^{-\Phi(q)(v-y)}\left\{W^{(q)'}(x - y) - \Phi(q)W^{(q)}(x - y)\right\}$$

$$\times \Pi(-du - v)dydv + \frac{\sigma^2}{2}\left(W^{(q)'}(x) - \Phi(q)W^{(q)}(x)\right)\delta_{(0,0,0)}(du, dv, dy),$$

where $\delta_{(0,0,0)}(du, dv, dy)$ is the Dirac delta measure.

Note that the second term on the right hand side corresponds to the event that ruin occurs by *creeping*. That is to say the process X enters $(-\infty, 0)$ continuously for the first time. It is clear from the formula in the statement of the theorem that creeping can only happen if and only if $\sigma^2 > 0$.

The use of scale functions has somewhat changed the landscape of ruin theory, offering access to a number of general results. See for example [5, 64, 69, 85], as well as the work mentioned in the next example.

Optimal Control: Suppose, as usual, that X is a spectrally negative Lévy process and consider the following optimal control problem

$$u(x) = \sup_\pi \mathbb{E}_x \left(\int_0^{\sigma^\pi} e^{-qt} dL_t^\pi \right), \tag{9}$$

where $x, q > 0$, $\sigma^\pi = \inf\{t > 0 : X_t - L_t^\pi < 0\}$ and $L^\pi = \{L_t^\pi : t \geq 0\}$ is the control process associated with the strategy π. The supremum is taken over all strategies π such that L^π is a non-decreasing, left-continuous adapted process which starts at zero and which does not allow the controlled process $X - L^\pi$ to enter $(-\infty, 0)$ as a consequence of one of its jumps. The solution to this control problem entails finding an expression for $u(x)$ and, if the supremum is attained by some strategy π^*, describing this optimal strategy.

This exemplary control problem originates from the work of de Finetti [42], again in the context of actuarial science. In the actuarial setting, X plays the role of a surplus process as usual and L^π is to be understood as a dividend payment. The objective then becomes to optimize the mean net present value of dividends paid until ruin of the aggregate surplus process occurs.

A very elegant solution to this problem has been obtained by Loeffen [75], following the work of Avram et al. [10], by imposing some additional assumptions on the underlying Lévy process, X. Loeffen's solution is intricately connected to certain analytical properties of the associated scale function $W^{(q)}$. By requiring that, for $x > 0$, $-\Pi(-\infty, -x)$ has a completely monotone density (recall that Π is the Lévy measure of X), it turns out that for each $q > 0$, the associated q-scale function, $W^{(q)}$, has a first derivative which is strictly convex on $(0, \infty)$. Moreover, the point $a^* := \inf\{a \geq 0 : W^{(q)\prime}(a) \leq W^{(q)\prime}(x) \text{ for all } x \geq 0\}$ provides a threshold from which an optimal strategy to (9) can be characterized; namely the strategy $L_t^* = (a^* \vee \overline{X}_t) - a^*$, $t \geq 0$. The aggregate process,

$$a^* + X_t - (a^* \vee \overline{X}_t), \ t \geq 0,$$

thus has the dynamics of the underlying Lévy process reflected at the barrier a^*. The value function of this strategy is given by

$$u(x) = \begin{cases} \frac{W^{(q)}(x)}{W^{(q)\prime}(a)}, & \text{for } 0 \leq x \leq a \\ x - a + \frac{W^{(q)}(a)}{W^{(q)\prime}(a)}, & \text{for } x > a. \end{cases} \tag{10}$$

The methodology that lead to the above solution has also been shown to have applicability in other related control problems. See for example [70, 71, 76–78].

Queuing Theory: The classical M/G/1 queue is described as follows. Customers arrive at a service desk according to a Poisson process with rate $\lambda > 0$ and join a queue. Customers have service times that are independent and identically distributed with common distribution F concentrated on $(0, \infty)$. Once served, they leave the queue.

The workload, $Y_t^{(x)}$, at each time $t \geq 0$, is defined to be the time it will take a customer who joins the back of the queue to reach the service desk when the workload at time zero is equal to $x \geq 0$. That is to say, $Y_t^{(x)}$ is the amount of processing time remaining in the queue at time t when a workload x already exists in the queue at time zero. If $\{X_t : t \geq 0\}$ is the difference of a pure linear drift of unit rate and a compound Poisson process, with rate λ and jump distribution F, then it is not difficult to show that when the current workload at time $t = 0$ is $x \geq 0$, then

$$Y_t^{(x)} = (x \vee \overline{X}_t) - X_t, \ t \geq 0.$$

Said another way, the workload has the dynamics of a spectrally positive Lévy process reflected at the origin.

Not surprisingly, many questions concerning the workload of the classical M/G/1 queue therefore boil down to issues regarding fluctuation theory of spectrally negative Lévy processes. A simple example concerns the issue of buffer overflow adjustment as follows. Suppose that there is a limited workload capacity in the queue, say $B > 0$. When the workload exceeds the buffer level B, the excess over level B is transferred elsewhere so that the adjusted workload process, say $\left\{Y_t^{(x,B)} : t \geq 0\right\}$, never exceeds the amount B. We are interested in the busy period for this M/G/1 queue with buffer overflow adjustment. That is to say,

$$\sigma_B^{(x)} := \inf\left\{t > 0 : Y_t^{(x,B)} = 0\right\}.$$

It is known that for all $q \geq 0$,

$$E\left(e^{-q\sigma_B^{(x)}}\right) = \frac{1 + q \int_0^{x-B} W^{(q)}(y)dy}{1 + q \int_0^B W^{(q)}(y)dy}.$$

See Pistorius [81] and Dube et al. [38] for more computations and applications in this vein.

Continuous State Branching Processes: A $[0, \infty]$-valued strong Markov process $Y = \{Y_t : t \geq 0\}$ with probabilities $\{P_x : x \geq 0\}$ is called a continuous-state branching process if it has paths that are right continuous with left limits and its law observes the following property. For any $t \geq 0$ and $y_1, y_2 \geq 0$, Y_t under $P_{y_1+y_2}$ is equal in law to the independent sum $Y_t^{(1)} + Y_t^{(2)}$ where the distribution of $Y_t^{(i)}$ is equal to that of Y_t under P_{y_i} for $i = 1, 2$.

It turns out that there is a one-to-one mapping between continuous state branching processes whose paths are not monotone and spectrally positive Lévy processes. Indeed, every continuous-state branching process Y may be written in the form

$$Y_t = X_{\theta_t \wedge \tau_0^-}, \ t \geq 0, \tag{11}$$

where X is a spectrally positive Lévy process, $\tau_0^- = \inf\{t > 0 : X_t < 0\}$ and

$$\theta_t = \inf\left\{s > 0 : \int_0^s \frac{du}{X_u} > t\right\}.$$

Conversely, for any given spectrally positive Lévy process X, the transformation on the right hand side of (11) defines a continuous-state branching process. (In fact the same bijection characterises all the continuous-state branching processes with monotone non-decreasing paths when X is replaced by a subordinator.)

A classic result due to Bingham [22] gives (under very mild conditions) the law of the maximum of the continuous-state branching process with non-monotone paths as follows. For all $x \geq y > 0$,

$$P_y(\sup_{s \geq 0} Y_s \leq x) = \frac{W(x - y)}{W(x)},$$

where W is the scale function associated with the underlying spectrally negative Lévy process in the representation (11). See Caballero et al. [26] and Kyprianou and Pardo [67] for further computations in this spirit and Lambert [73] for further applications in population biology.

Fragmentation Processes: A homogenous mass fragmentation process is a Markov process $\mathbf{X} := \{\mathbf{X}(t) : t \geq 0\}$, where $\mathbf{X}(t) = (X_1(t), X_2(t), \cdots)$, that takes values in

$$\mathscr{S} := \left\{\mathbf{s} = (s_1, s_2, \cdots) : s_1 \geq s_2 \geq \cdots \geq 0, \sum_{i=1}^{\infty} s_i = 1\right\}.$$

Moreover $\mathbf{X}$ possesses the fragmentation property as follows. Suppose that for each $\mathbf{s} \in \mathscr{S}$, $\mathbb{P}_\mathbf{s}$ denotes the law of $\mathbf{X}$ with $\mathbf{X}(0) = \mathbf{s}$. Given, for $t \geq 0$, that $\mathbf{X}(t) = (s_1, s_2, \cdots) \in \mathscr{S}$, then for all $u > 0$, $\mathbf{X}(t + u)$ has the same law as the variable obtained by ranking in decreasing order the elements contained in the sequences $\mathbf{X}^{(1)}(u), \mathbf{X}^{(2)}(u), \cdots$, where the latter are independent, random mass partitions with values in $\mathscr{S}$ having the same distribution as $\mathbf{X}(u)$ under $\mathbb{P}_{(s_1,0,0,\cdots)}$, $\mathbb{P}_{(s_2,0,0,\cdots)}$, $\cdots$ respectively. The process $\mathbf{X}$ is homogenous in the sense that, for every $r > 0$, the law of $\{r\mathbf{X}(t) : t \geq 0\}$ under $\mathbb{P}_{(1,0,0,\cdots)}$ is $\mathbb{P}_{(r,0,0,\cdots)}$.

Similarly to branching processes, fragmentation processes have embedded genealogies in the sense that a block at time t is a descendent from a block at time $s < t$ if it has fragmented from it. It is possible to formulate the notion of the evolution of such a genealogical line of descent chosen "uniformly at random." To avoid a long, detailed exposition, we refrain from providing the technical details here, and mention instead that if $\{X^*(t) : t \geq 0\}$ is the sequence of embedded fragments along the aforesaid uniformly chosen genealogical line of descent, then it turns out that $\{-\log X^*(t) : t \geq 0\}$ is a subordinator. We suppose that, for $\theta \geq 0$, $\phi(\theta) = \mathbb{E}_{(1,0,0,\dots)}(X^*(t)^\theta)$ is the Laplace exponent of this subordinator.

Knobloch and Kyprianou [55] show that for appropriate values of $c, p \geq 0$,

$$M_t := \sum_{i \in \mathscr{I}_t^c} e^{-\psi(p)t} X_i(t) W^{(\psi(p))}(ct + \log X_i(t)), \ t \geq 0, \tag{12}$$

is a martingale, where for $\theta \geq 0$, $\psi(\theta) = c\theta - \phi(\theta)$ is the Laplace exponent of a spectrally negative Lévy process, $W^{(q)}$ is the q-scale function with respect to ψ and $\mathscr{I}_t^c$ is the set of indices of fragments at time t whose genealogical line of descent has the property that, for all $s \leq t$, its ancestral fragment at time s is larger than e^{-cs}. One may think of the sum as being over a "thinned" version of the original fragmentation process. That is to say, an adjustment of the original process in which blocks are removed if, at time $t \geq 0$, they become smaller than e^{-ct}. Removed blocks may therefore no longer contribute fragmented mass to the on-going process of fragmentation.

Analysis of this martingale in [55], in particular making use of known properties of scale functions, allows the authors to deduce that, under mild conditions, there exists a unique constant $p^* > 0$ such that whenever $c > \phi'(p^*)$

$$\sup_{i \in \mathscr{I}_t^c} -\frac{\log X_i(t)}{t} \to \phi'(p^*), \tag{13}$$

as $t \uparrow \infty$ on the event that the index set $\mathscr{I}_t^c$ remains non-empty for all $t \geq 0$ (i.e. the thinned process survives), which itself occurs with positive probability. This result is of interest when one compares it against the growth of the largest fragment without restriction of the index set. In that case it is known that

$$-\frac{\log X_i(t)}{t} \to \phi'(p^*),$$

as $t \uparrow \infty$. Intuitively speaking (13) says that the thinned fragmentation process will either become eradicated (all blocks are removed in a finite time) or the original fragmentation process will survive the thinning procedure and the decay rate of the largest block is unaffected.

A more elaborate martingale, similar in nature to (12) and also built using a scale function, was used by Krell [61] for a more detailed analysis of different rates of fragmentation decay that can occur in the process. In particular, a fractal analysis of the different decay rates is possible.

Lévy Processes Conditioned to Stay Positive: What is important in many of the examples above is the behaviour of the underlying spectrally negative Lévy process as it exits a half line. In contrast, one may consider the behaviour of such Lévy processes conditioned never to exit a half line. This is of equal practical value from a modelling point of view, as well as having the additional curiosity that the conditioning event can occur with probability zero.

Take, as usual, X to be a spectrally negative Lévy process and recall the definition (8) of τ_0^-. Assume that $\psi'(0+) \geq 0$, then it is known that for all $t \geq 0$ and $x, y > 0$,

$$P_x^\uparrow (X_t \in dy) = \lim_{q \downarrow 0} P_x \left(X_t \in dy, \, t < \mathbf{e}/q | \tau_0^- > \mathbf{e}/q \right), \qquad (14)$$

where $\mathbf{e}$ is an independent and exponentially distributed random variable with unit mean, defines the semigroup of a conservative strong Markov process which can be meaningfully called a *spectrally negative Lévy process conditioned to stay positive*. Moreover, it turns out to be the case that the laws $\{ P_x^\uparrow : x > 0 \}$ can be described through the laws $\{ P_x : x > 0 \}$ via a classical Doob h-transform. Indeed, for all $A \in \mathscr{F}_t$ and $x, t > 0$,

$$P_x^\uparrow (A) = E_x \left(\frac{W(X_t)}{W(x)} \mathbf{1}_{\{ A, \, \tau_0^- > t \}} \right).$$

Hence a significant amount of probabilistic information concerning this conditioning is captured by the scale function. See Chaumont and Doney [30] for a complete overview. In a similar spirit Chaumont [28,29] also shows that scale functions can be used to describe the law of a Lévy process conditioned to hit the origin continuously in a finite time. Later on in this review we shall see another example of conditioning spectrally negative Lévy processes due to Lambert [72], which again involves the scale function in a similar spirit.

2   The General Theory of Scale Functions

2.1   Some Additional Facts About Spectrally Negative Lévy Processes

Let us return briefly to the family of spectrally negative Lévy processes and remind ourselves of some additional deeper properties thereof which, as we shall see, are closely intertwined with the properties of scale functions.

Path Variation. Over and above the assumption of spectral negativity in (1), the conditions

$$\int_{(-1,0)} |x| \Pi(dx) < \infty \text{ and } \sigma = 0,$$

are necessary and sufficient for X to have paths of bounded variation. In that case X may necessarily be decomposed uniquely in the form

$$X_t = \delta t - S_t, \, t \geq 0, \qquad (15)$$

where $\delta > 0$ and $\{S_t : t \geq 0\}$ is a pure jump subordinator.

Regularity of the Half Line: Recall that *irregularity of* $(-\infty, 0)$ *for* 0 for X means that $P(\tau_0^- > 0) = 1$ where

$$\tau_0^- = \inf \{t > 0 : X_t < 0\}.$$

Thanks to Blumenthal's zero-one law, the only alternative to irregularity of $(-\infty, 0)$ for 0 is *regularity of* $(-\infty, 0)$ *for* 0 meaning that $P(\tau_0^- > 0) = 0$. Roughly speaking one may say in words that a spectrally negative Lévy process enters the open lower half line a.s. immediately from 0 if and only if it has paths of unbounded variation, otherwise it takes an a.s. positive amount of time before visiting the open lower half line. In contrast, for all spectrally negative Lévy processes we have $\mathbb{P}(\tau_0^+ = 0) = 1$, where

$$\tau_0^+ = \inf \{t > 0 : X_t > 0\}.$$

That is to say, there is always regularity of $(0, \infty)$ for 0.

Creeping. When a spectrally negative Lévy process issued from $x > 0$ enters $(-\infty, 0)$ for the first time, it may do so either by a jump or continuously. That is to say, on the event $\{\tau_0^- < \infty\}$, either $X_{\tau_0^-} = 0$ or $X_{\tau_0^-} < 0$. If the former happens with positive probability it is said that the process *creeps* downwards. We are deliberately vague here about the initial value $x > 0$ under which creeping occurs as it turns out that if $P_y(X_{\tau_0^-} = 0) > 0$ for some $y > 0$, then $P_x(X_{\tau_0^-} = 0) > 0$ for all $x > 0$. It is known that the only spectrally negative Lévy processes which can creep downwards are those processes which have a Gaussian component. Note that, thanks to the fact that there are no positive jumps, a spectrally negative Lévy process will always a.s. creep upwards over any level above its initial position, providing the path reaches that level. That is to say, for all $a \geq x$, $P_x(X_{\tau_a^+} = a; \tau_a^+ < \infty) = P_x(\tau_a^+ < \infty)$ where

$$\tau_a^+ = \inf \{t \geq 0 : X_t > a\}.$$

Exponential Change of Measure. The Laplace exponent also provides a natural instrument with which one may construct an "exponential change of measure" on X in the sprit of the classical Cameron–Martin–Girsanov transformation; a technique which plays a central role throughout this text. The equality (2) allows for a Girsanov-type change of measure to be defined, namely via

$$\left. \frac{\mathrm{d}P_x^c}{\mathrm{d}P_x} \right|_{\mathscr{F}_t} = e^{c(X_t - x) - \psi(c)t}, \, t \geq 0, \qquad (16)$$

for any c such that $|\psi(c)| < \infty$. Note that it is a straightforward exercise to show that the right hand side above is a martingale thanks to the fact that X has stationary independent increments together with (2). Moreover, the absolute continuity implies that under this change of measure, X remains within the class of spectrally negative processes and the Laplace exponent of X under P_x^c is given by

$$\psi_c(\theta) = \psi(\theta + c) - \psi(c), \tag{17}$$

for $\theta \geq -c$. If Π_c denotes the Lévy measure of (X, P^c), then it is straightforward to deduce from (17) that

$$\Pi_c(\mathrm{d}x) = \mathrm{e}^{cx} \Pi(\mathrm{d}x),$$

for $x < 0$.

The Wiener–Hopf Factorization. Suppose that $\underline{X}_t := \inf_{s \leq t} X_s$ and $\mathbf{e}_q$ is an independent and exponentially distributed random variable with rate $q > 0$, then the variables $X_{\mathbf{e}_q} - \underline{X}_{\mathbf{e}_q}$ and $-\underline{X}_{\mathbf{e}_q}$ are independent. The fundamental but simple concept of duality for Lévy processes, which follows as a direct consequence of stationary and independent increments and càdlàg paths states that $\{X_t - X_{(t-s)-} : s \leq t\}$ is equal in law to $\{X_s : s \leq t\}$. This in turn implies that $X_t - \underline{X}_t$ is equal in distribution to $\overline{X}_t := \sup_{s \leq t} X_s$. It follows that for all $\theta \in \mathbb{R}$,

$$E\left(\mathrm{e}^{\mathrm{i}\theta X_{\mathbf{e}_q}}\right) = E\left(\mathrm{e}^{\mathrm{i}\theta \overline{X}_{\mathbf{e}_q}}\right) E\left(\mathrm{e}^{\mathrm{i}\theta \underline{X}_{\mathbf{e}_q}}\right). \tag{18}$$

This identity is known as the Wiener–Hopf factorization. Note that the reasoning thus far applies to the case that X is a general Lévy process. However, when X is a spectrally negative Lévy process, it is possible to express the right hand side above in a more convenient analytical form. Let $a > 0$, since $t \wedge \tau_a^+$ is a bounded stopping time and $X_{t \wedge \tau_a^+} \leq a$, it follows from Doob's Optional Stopping Theorem applied to the exponential martingale in (16) that

$$E\left(\mathrm{e}^{\Phi(q)X_{t \wedge \tau_a^+} - q(t \wedge \tau_a^+)}\right) = 1.$$

By dominated convergence and the fact that $X_{\tau_a^+} = a$ on $\tau_a^+ < \infty$ we have,

$$P\left(\overline{X}_{\mathbf{e}_q} > a)\right) = P\left(\tau_a^+ < \mathbf{e}_q\right)$$

$$= E\left(\mathrm{e}^{-q\tau_a^+} \mathbf{1}_{(\tau_a^+ < \infty)}\right)$$

$$= \mathrm{e}^{-\Phi(q)a}. \tag{19}$$

The conclusion of the above series of equalities is that the quantity $\overline{X}_{\mathbf{e}_q}$ is exponentially distributed with parameter $\Phi(q)$ and therefore for all $\theta \in \mathbb{R}$,

$$E\left(e^{i\theta \overline{X}_{e_q}}\right) = \frac{\Phi(q)}{\Phi(q) - i\theta}. \tag{20}$$

It follows from the factorization (18) that for all $\theta \in \mathbb{R}$

$$E\left(e^{i\theta \underline{X}_{e_q}}\right) = \frac{q}{\Phi(q)} \frac{\Phi(q) - i\theta}{q + \Psi(\theta)}. \tag{21}$$

As we shall see later on, a number of features described above concerning the Wiener–Hopf factorization, including the previous identity, play an instrumental role in the theory and application of scale functions.

2.2 Existence of Scale Functions

We are now ready to show the existence of scale functions. That is to say, we will show that the function $(\psi(\beta) - q)^{-1}$ is genuinely a Laplace transform in β. This will largely be done through the Wiener–Hopf factorization in combination with an exponential change of measure.

Theorem 2.1. *For all spectrally negative Lévy processes, q-scale functions exist for all $q \geq 0$.*

Proof. First assume that $\psi'(0+) > 0$. With a pre-emptive choice of notation, define the function

$$W(x) = \frac{1}{\psi'(0+)} P_x(\underline{X}_\infty \geq 0). \tag{22}$$

Clearly $W(x) = 0$ for $x < 0$, it is right-continuous since it is also proportional to the distribution function $P(-\underline{X}_\infty \leq x)$. Integration by parts shows that, on the one hand,

$$
\begin{aligned}
\int_0^\infty e^{-\beta x} W(x) dx &= \frac{1}{\psi'(0+)} \int_0^\infty e^{-\beta x} P(-\underline{X}_\infty \leq x) \, dx \\
&= \frac{1}{\psi'(0+)\beta} \int_{[0,\infty)} e^{-\beta x} \, P(-\underline{X}_\infty \in dx) \\
&= \frac{1}{\psi'(0+)\beta} E\left(e^{\beta \underline{X}_\infty}\right).
\end{aligned} \tag{23}
$$

On the other hand, taking limits as $q \downarrow 0$ in (21) gives us the identity

$$E\left(e^{i\theta \underline{X}_\infty}\right) = -\psi'(0+)\frac{i\theta}{\Psi(\theta)},$$

where, thanks to the assumption that $\psi'(0+) > 0$, we have used the fact that $\Phi(0) = 0$ and hence

$$\lim_{q\downarrow 0} \frac{q}{\Phi(q)} = \lim_{q\downarrow 0} \frac{\psi(\Phi(q))}{\Phi(q)} = \lim_{\theta\downarrow 0} \frac{\psi(\theta)}{\theta} = \psi'(0+).$$

A straightforward argument using analytical extension now gives us the identity

$$E\left(e^{\beta X_\infty}\right) = \frac{\psi'(0+)\beta}{\psi(\beta)}. \tag{24}$$

Combining (23) with (24) gives us (4) as required for the case $q = 0$ and $\psi'(0+) > 0$.

Next we deal with the case where $q > 0$ or $q = 0$ *and* $\psi'(0+) < 0$. To this end, again making use of a pre-emptive choice of notation, let us define

$$W^{(q)}(x) = e^{\Phi(q)x}W_{\Phi(q)}(x), \tag{25}$$

where $W_{\Phi(q)}$ plays the role of W but for the process $(X, P^{\Phi(q)})$. Note in particular that by (17) the latter process has Laplace exponent

$$\psi_{\Phi(q)}(\theta) = \psi(\theta + \Phi(q)) - q, \tag{26}$$

for $\theta \geq 0$, and hence $\psi'_{\Phi(q)}(0+) = \psi'(\Phi(q)) > 0$, which ensures that $W_{\Phi(q)}$ is well defined by the previous part of the proof. Taking Laplace transforms we have for $\beta > \Phi(q)$,

$$\int_0^\infty e^{-\beta x} W^{(q)}(x)\mathrm{d}x = \int_0^\infty e^{-(\beta-\Phi(q))x} W_{\Phi(q)}(x)\mathrm{d}x$$

$$= \frac{1}{\psi_{\Phi(q)}(\beta - \Phi(q))}$$

$$= \frac{1}{\psi(\beta) - q},$$

thus completing the proof for the case $q > 0$ or $q = 0$ and $\psi'(0+) < 0$.

Finally, the case that $q = 0$ and $\psi'(0+) = 0$ can be dealt with as follows. Since $W_{\Phi(q)}(x)$ is an increasing function, we may also treat it as a distribution function of a measure which we also, as an abuse of notation, call $W_{\Phi(q)}$. Integrating by parts thus gives us for $\beta > 0$,

$$\int_{[0,\infty)} e^{-\beta x} W_{\Phi(q)}(\mathrm{d}x) = \frac{\beta}{\psi_{\Phi(q)}(\beta)}. \tag{27}$$

Note that the assumption $\psi'(0+) = 0$, implies that $\Phi(0) = 0$, and hence for $\theta \geq 0$,

$$\lim_{q \downarrow 0} \psi_{\Phi(q)}(\theta) = \lim_{q \downarrow 0}[\psi(\theta + \Phi(q)) - q] = \psi(\theta).$$

One may appeal to the Extended Continuity Theorem for Laplace Transforms, see [40] Theorem XIII.1.2a, and (27) to deduce that since

$$\lim_{q \downarrow 0} \int_{[0,\infty)} e^{-\beta x} W_{\Phi(q)}(\mathrm{d}x) = \frac{\beta}{\psi(\beta)},$$

then there exists a measure W^* such that $W^*(x) := W^*[0,x] = \lim_{q \downarrow 0} W_{\Phi(q)}(x)$ and

$$\int_{[0,\infty)} e^{-\beta x} W^*(\mathrm{d}x) = \frac{\beta}{\psi(\beta)}.$$

Integration by parts shows that the right-continuous distribution,

$$W(x) := W^*[0,x]$$

satisfies

$$\int_0^{\infty} e^{-\beta x} W(x)\, \mathrm{d}x = \frac{1}{\psi(\beta)},$$

for $\beta > 0$ as required.   □

Let us return to Theorem 1.2 and show that, now that we have a slightly clearer understanding of what a scale function is, a straightforward proof of the aforementioned identity can be given.

Proof (of Theorem 1.2). First we deal with the case that $q = 0$ and $\psi'(0+) > 0$ as in the previous proof. Since we have identified $W(x) = P_x\left(\underline{X}_\infty \geq 0\right)/\psi'(0+)$, a simple argument using the law of total probability and the Strong Markov Property now yields for $x \in [0,a]$

$$P_x(\underline{X}_\infty \geq 0)$$
$$= E_x\left(P_x\left(\underline{X}_\infty > 0 \,|\, \mathscr{F}_{\tau_a^+}\right)\right)$$
$$= E_x\left(\mathbf{1}_{(\tau_a^+ < \tau_0^-)} P_a\left(\underline{X}_\infty \geq 0\right)\right) + E_x\left(\mathbf{1}_{(\tau_a^+ > \tau_0^-)} P_{X_{\tau_0^-}}\left(\underline{X}_\infty \geq 0\right)\right).$$
$$\tag{28}$$

The first term on the right hand side above is equal to

$$P_a\left(\underline{X}_\infty \geq 0\right) P_x\left(\tau_a^+ < \tau_0^-\right).$$

The second term on the right hand side of (28) is more complicated to handle but none the less is always equal to zero. To see why, first suppose that X has no Gaussian component. In that case it cannot creep downwards and hence $X_{\tau_0^-} < 0$

and the claim follows by virtue of the fact that $P_x(\underline{X}_\infty \geq 0) = 0$ for $x < 0$. If, on the other hand, X has a Gaussian component, the previous argument still applies on the event $\{X_{\tau_0^-} < 0\}$, however we must also consider the event $\{X_{\tau_0^-} = 0\}$. In that case we note that regularity of $(-\infty, 0)$ for 0 for X implies that $P(\underline{X}_\infty \geq 0) = 0$. Returning to (28) we may now deduce that

$$P_x\left(\tau_a^+ < \tau_0^-\right) = \frac{W(x)}{W(a)}, \tag{29}$$

and clearly the same equality holds even when $x < 0$ as both left and right hand side are identically equal to zero.

Next we deal with the case $q > 0$. Making use of (2) in a, by now, familiar way, and recalling that $X_{\tau_a^+} = a$, we have that

$$
\begin{aligned}
E_x\left(e^{-q\tau_a^+} \mathbf{1}_{(\tau_a^+ < \tau_0^-)}\right) &= E_x\left(e^{\Phi(q)(X_{\tau_a^+} - x) - q\tau_a^+} \mathbf{1}_{(\tau_a^+ < \tau_0^-)}\right) e^{-\Phi(q)(a-x)} \\
&= e^{-\Phi(q)(a-x)} P_x^{\Phi(q)}\left(\tau_a^+ < \tau_0^-\right) \\
&= e^{-\Phi(q)(a-x)} \frac{W_{\Phi(q)}(x)}{W_{\Phi(q)}(a)} \\
&= \frac{W^{(q)}(x)}{W^{(q)}(a)}.
\end{aligned}
$$

Finally, to deal with the case that $q = 0$ and $\psi'(0+) \leq 0$, one needs only to take limits as $q \downarrow 0$ in the above identity, making use of monotone convergence on the left hand side and continuity in q on the right hand side thanks to the Continuity Theorem for Laplace transforms. □

Before moving on to looking at the intimate relation between scale functions and the so-called excursion measure associated with X, let us make some further remarks about the scale function and its relation to the Wiener–Hopf factorization. It is clear from the proof of Theorem 2.1 that the very definition of a scale function is embedded in the Wiener–Hopf factorization. The following corollary to the aforementioned theorem reinforces this idea.

Corollary 2.2. *For $x \geq 0$,*

$$P(-\underline{X}_{e_q} \in dx) = \frac{q}{\Phi(q)} W^{(q)}(dx) - qW^{(q)}(x)dx. \tag{30}$$

Proof. A straightforward argument using analytical extension allows us to re-write the Wiener–Hopf factor (21) as a Laplace transform,

$$E\left(e^{\beta \underline{X}_{e_q}}\right) = \frac{q}{\Phi(q)} \frac{\beta - \Phi(q)}{\psi(\beta) - q},$$

for $\beta \geq 0$. Note in particular, when $\beta = \Phi(q)$ the right hand side is understood in the limiting sense using L'Hôpital's rule. The identity (30) now follows directly from the Laplace transform (4) and an integration by parts. $\square$

2.3   Scale Functions and the Excursion Measure

Section 2.2 shows that there is an intimate relationship between scale functions and the Wiener–Hopf factorization. The Wiener–Hopf factorization can itself be seen as a distributional consequence of a decomposition of the path of any Lévy process into excursions from its running maximum, so-called *excursion theory*; see for example the presentation in Greenwood and Pitman [46], Bertoin [15] and Kyprianou [66]. Let us briefly spend some time reviewing the theory of excursions within the context of spectrally negative Lévy processes and thereafter we will show the connection between scale functions and a key object which plays a central role in the latter known as the *excursion measure*.

The basic idea of excursion theory for a Lévy process is to provide a structured mathematical description of the successive sections of its trajectory which make up sojourns, or excursions, from its previous maximum. In order to do this, we need to introduce a new time-scale which will help us locate the times at which X creates new maxima. Specifically we are interested in a (random) function of original time, or clock, say $L = \{L_t : t \geq 0\}$, such that L increases precisely at times when X creates a new maximum; namely the times $\{t : X_t = \overline{X}_t\}$. Moreover, L should have a regenerative property in that if T is any $\mathbb{F}$-stopping time such that $\overline{X}_T = X_T$ on $\{T < \infty\}$ then $\{(X_{T+t} - X_T, L_{T+t} - L_T) : t \geq 0\}$ has the same law as $\{(X_t, L_t) : t \geq 0\}$ under P. The process L is referred to as *local time at the maximum*.

It turns out that in the very special case of spectrally negative Lévy processes, there is a natural choice of L which is simply $L = \overline{X}$. Indeed it is straightforward to verity that $\overline{X}$ has the required regenerative property. We shall henceforth proceed with this choice of L. Now define

$$L_t^{-1} = \inf \left\{ s > 0 : L_s > t \right\}, \quad t \geq 0,$$

Note that in the above definition, t is a local time and L_t^{-1} is the real time which is required to accumulate t units of local time. On account of the fact that $L = \overline{X}$, it is also the case that L_t^{-1} is the first passage time of the process X above level t. In the notation of the previous section, $L_t^{-1} = \tau_t^+$.

Standard theory tells us that, for all $t > 0$, both L_t^{-1} and L_{t-}^{-1} are stopping times. Moreover, it is quite clear that whenever $\Delta L_t^{-1} := L_t^{-1} - L_t^{-1} > 0$, the process X experiences an excursion from its previous maximum $\overline{X} = t$. For such local times the associated excursion shall be written as

$$\epsilon_t = \left\{ \epsilon_t(s) := X_{L_t^{-1}} - X_{L_{t-}^{-1}+s} : 0 < s \leq \Delta L_t^{-1} \right\}.$$

For local times $t > 0$ such that $\Delta L_t^{-1} = 0$ we define $\epsilon_t = \partial$ where ∂ is some isolated state. For each $t > 0$ such that $\Delta L_t^{-1} > 0$, ϵ_t takes values in the space of excursions, $\mathcal{E}$, the space of real valued, right-continuous paths with left limits, killed at the first hitting time of $(-\infty, 0]$. We suppose that $\mathcal{E}$ is endowed with the sigma-algebra generated by the coordinate maps, say $\mathcal{G}$.

The key feature of excursion theory in the current setting is that

$$\{(t, \epsilon_t) : t \geq 0 \text{ and } \epsilon_t \neq \partial\}$$

is a Poisson point process with values in $(0, \infty) \times \mathcal{E}$ and intensity measure $dt \times dn$, where n is a measure on the space $(\mathcal{E}, \mathcal{G})$.

Let us now return to the relationship between excursion theory and the scale function. First write ζ for the lifetime and $\bar{\epsilon}$ for the height of the generic excursion $\epsilon \in \mathcal{E}$. That is to say, for $\epsilon \in \mathcal{E}$,

$$\zeta = \inf\{s > 0 : \epsilon(s) \leq 0\} \text{ and } \bar{\epsilon} = \sup\{\epsilon(s) : s \leq \zeta\}.$$

Choose $a > x \geq 0$. Recall that $L = \overline{X}$ we have that $L_{\tau_{a-x}^+} = \overline{X}_{\tau_{a-x}^+} = a - x$. Using this it is not difficult to see that

$$\left\{\underline{X}_{\tau_{a-x}^+} \geq -x\right\} = \{\forall t \leq a - x \text{ and } \epsilon_t \neq \partial, \bar{\epsilon}_t \leq t + x\}.$$

Let N be the Poisson random measure associated with the Poisson point process of excursions; that is to say for all $dt \times dn$-measurable sets $B \in [0, \infty) \times \mathcal{E}$, $N(B) = \text{card}\{(t, \epsilon_t) \in B, t > 0\}$. With the help of (29) we have that

$$\begin{aligned}
\frac{W(x)}{W(a)} &= P_x\left(\underline{X}_{\tau_a^+} \geq 0\right) \\
&= P\left(\underline{X}_{\tau_{a-x}^+} \geq -x\right) \\
&= P(\forall t \leq a - x \text{ and } \epsilon_t \neq \partial, \bar{\epsilon}_t \leq t + x) \\
&= P(N(A) = 0),
\end{aligned}$$

where $A = \{(t, \epsilon_t) : t \leq a - x \text{ and } \bar{\epsilon}_t > t + x\}$. Since $N(A)$ is Poisson distributed with parameter $\int \mathbf{1}_A \, dt \, n\,(d\epsilon) = \int_0^{a-x} n\,(\bar{\epsilon} > t + x) \, dt = \int_x^a n\,(\bar{\epsilon} > t) \, dt$ we have that

$$\frac{W(x)}{W(a)} = \exp\left\{-\int_x^a n\,(\bar{\epsilon} > t) \, dt\right\}. \tag{31}$$

This is a fundamental representation of the scale function W which we shall return to in later sections. However, for now, it leads us immediately to the following analytical conclusion.

Lemma 2.3. *For any $q \geq 0$, the scale function $W^{(q)}$ is continuous, almost everywhere differentiable and strictly increasing.*

Proof. Assume that $q = 0$. Since a may be chosen arbitrarily large, continuity and strict monotonicity follow from (31). Moreover, we also see from (31) that the left and right first derivatives exist and are given by

$$W'_-(x) = n(\bar{\epsilon} \geq x) W(x) \text{ and } W'_+(x) = n(\bar{\epsilon} > x) W(x). \tag{32}$$

Since there can be at most a countable number of points for which the monotone function $n(\bar{\epsilon} > x)$ has discontinuities, it follows that W is almost everywhere differentiable.

From the relation (25) we know that

$$W^{(q)}(x) = e^{\Phi(q)x} W_{\Phi(q)}(x), \tag{33}$$

and hence the properties of continuity, almost everywhere differentiability and strict monotonicity carry over to the case $q > 0$. $\qquad\square$

As an abuse of notation, let us write $W^{(q)} \in C^1(0, \infty)$ to mean that the restriction of $W^{(q)}$ to $(0, \infty)$ belongs to $C^1(0, \infty)$. Then the proof of the previous lemma indicates that $W \subset C^1(0, \infty)$ as soon as the measure $n(\bar{\epsilon} \in \mathrm{d}x)$ has no atoms. It is possible to combine this observation together with other analytical and probabilistic facts to recover necessary and sufficient conditions for first degree smoothness of W.

Lemma 2.4. *For each $q \geq 0$, the scale function $W^{(q)}$ belongs to $C^1(0, \infty)$ if and only if at least one of the following three criteria holds,*

(i) $\sigma \neq 0$
(ii) $\int_{(-1,0)} |x| \Pi(\mathrm{d}x) = \infty$
(iii) $\overline{\Pi}(x) := \Pi(-\infty, -x)$ *is continuous.*

Proof. Firstly note that it suffices to consider the case that $q = 0$ and $\psi'(0+) \geq 0$, i.e. the case that the Lévy process oscillates or drifts to $+\infty$. Indeed, in the case that $\psi'(0+) < 0$ or $q > 0$, one recalls that $\Phi(0) > 0$, respectively $\Phi(q) > 0$, and we infer smoothness from (25), where $W_{\Phi(0)}$, respectively $W_{\Phi(q)}$, is the scale function of a spectrally negative Lévy process which drifts to $+\infty$.

If it is the case that there exists an $x > 0$, such that $n(\bar{\epsilon} = x) > 0$, then one may consider the probability law $n(\cdot|\bar{\epsilon} = x) = n(\cdot, \bar{\epsilon} = x)/n(\bar{\epsilon} = x)$. Assume now that X has paths of unbounded variation, that is to say (i) or (ii) holds, and recall that in this case 0 is regular for $(-\infty, 0)$. If we apply the strong Markov property under $n(\cdot|\bar{\epsilon} = x)$ to the excursion at the time $T_x := \inf\{t > 0 : \epsilon(t) = x\}$, the aforementioned regularity of X implies that (x, ∞) is regular for $\{x\}$ for the process ϵ and therefore that $\bar{\epsilon} > x$, under $n(\cdot|\bar{\epsilon} = x)$. However this constitutes a contradiction. It thus follows that $n(\bar{\epsilon} = x) = 0$ for all $x > 0$.

Now assume that X has paths of bounded variation. It is known from [82] and Proposition 5 in [106] that when X is of bounded variation the excursion measure of X reflected at its supremum begins with a jump and can be described by the formula

$$n\left(F\left(\epsilon(s), 0 \leq s \leq \varsigma\right)\right) = \frac{1}{\delta} \int_{-\infty}^{0} \Pi(dx)\widehat{E}_{-x}\left(F(X_s, 0 \leq s \leq \tau_0^-)\right),$$

where F is any nonnegative measurable functional on the space of cadlag paths, $\widehat{E}_x$ denotes the law of the dual Lévy process $\widehat{X} = -X$ and $\tau_x^- = \inf\{s > 0 : X_s < x\}$, $x \in \mathbb{R}$. Recall that $\tau_z^+ = \inf\{s > 0 : X_s > z\}$, $z \in \mathbb{R}$. Hence, it follows that

$$n\left(\bar{\epsilon} > z\right) = \frac{1}{\delta} \int_{(-\infty,0)} \Pi(dx)\widehat{P}_{-x}\left(\sup_{0 \leq s \leq \tau_0^-} X_s > z\right)$$

$$= \frac{1}{\delta}\Pi(-\infty, -z) + \frac{1}{\delta} \int_{[-z,0)} \Pi(dx)\widehat{P}_{-x}\left(\tau_z^+ < \tau_0^-\right)$$

$$= \frac{1}{\delta}\Pi(-\infty, -z) + \frac{1}{\delta} \int_{[-z,0)} \Pi(dx)\widehat{P}\left(\tau_{z+x}^+ < \tau_x^-\right)$$

$$= \frac{1}{\delta}\Pi(-\infty, -z) + \frac{1}{\delta} \int_{[-z,0)} \Pi(dx)P\left(\tau_{-x-z}^- < \tau_{-x}^+\right)$$

$$= \frac{1}{\delta}\Pi(-\infty, -z) + \frac{1}{\delta} \int_{[-z,0)} \Pi(dx)\left(1 - \frac{W(z+x)}{W(z)}\right),$$

where in the last equality we have used Theorem 1.2 with $q = 0$. From this it follows that

$$n\left(\bar{\epsilon} = z\right) = \frac{1}{\delta}\Pi(\{-z\})\frac{W(0)}{W(z)},$$

and hence that $n\left(\bar{\epsilon} = z\right) = 0$ (which in turn leads to the $C^1(0, \infty)$ property of the scale function) as soon as Π has no atoms. That is to say $\overline{\Pi}$ is continuous. $\qquad\square$

The proof of the above lemma in the bounded variation case gives us a little more information about non-differentiability in the case that Π has atoms.

Corollary 2.5. *When X has paths of bounded variation the scale function $W^{(q)}$ does not possess a derivative at $x > 0$ (for all $q \geq 0$) if and only if Π has an atom at $-x$. In particular, if Π has a finite number of atoms supported by the set $\{-x_1, \cdots, -x_n\}$ then, for all $q \geq 0$, $W^{(q)} \in C^1((0, \infty)\backslash\{x_1, \cdots, x_n\})$.*

The above results give us our first analytical impressions on the smoothness of $W^{(q)}$. In Sect. 3 we shall explore these properties in greater detail.

2.4 Scale Functions and the Descending Ladder Height Process

In a similar spirit to the previous section, it is also possible to construct an excursion theory for spectrally negative Lévy processes away from their infimum. Indeed it is well known, and easy to prove, that the process X reflected in its past infimum $X - \underline{X} := \{X_t - \underline{X}_t : t \geq 0\}$, where $\underline{X}_t := \inf_{s \leq t} X_s$, is a strong Markov process with state space $[0, \infty)$. Following standard theory of Markov local times (cf. Chap. IV of [15]), it is possible to construct a local time at zero for $X - \underline{X}$ which we henceforth refer to as $\widehat{L} = \{\widehat{L}_t : t \geq 0\}$. Its right-continuous inverse process, $\widehat{L}^{-1} := \{\widehat{L}_t^{-1} : t \geq 0\}$ where $\widehat{L}_t^{-1} = \inf\{s > 0 : \widehat{L}_s > t\}$, is a (possibly killed) subordinator. Sampling X at $\widehat{L}^{-1}$ we recover the points of local minima of X. If we define $\widehat{H}_t = X_{\widehat{L}_t^{-1}}$ when $\widehat{L}_t^{-1} < \infty$, with $\widehat{H}_t = \infty$ otherwise, then it is known that the process $\widehat{H} = \{\widehat{H}_t : t \geq 0\}$ is a (possibly killed) subordinator. The latter is called the *descending ladder height process*.

The starting point for the relationship between the descending ladder height process and scale functions is given by the following factorization of the Laplace exponent of X,

$$\psi(\theta) = (\theta - \Phi(0))\phi(\theta), \qquad \theta \geq 0, \tag{34}$$

where ψ, is the Laplace exponent of a (possibly killed) subordinator. See e.g. Sect. 6.5.2 in [66]. This can be seen by recalling that $\Phi(0)$ is a root of the equation $\psi(\theta) = 0$ and then factoring out $(\theta - \Phi(0))$ from ψ by using integration by parts to deduce that the term $\phi(\theta)$ must necessarily take the form

$$\phi(\theta) = \kappa + \xi\theta + \int_{(0,\infty)} (1 - e^{-\theta x})\Upsilon(dx), \tag{35}$$

where $\kappa, \xi \geq 0$ and Υ is a measure concentrated on $(0, \infty)$ which satisfies $\int_{(0,\infty)} (1 \wedge x)\Upsilon(dx) < \infty$. Indeed, it turns out that

$$\Upsilon(x, \infty) = e^{\Phi(0)x} \int_x^\infty e^{-\Phi(0)u} \Pi(-\infty, -u)du \quad \text{for } x > 0, \tag{36}$$

$\xi = \sigma^2/2$ and $\kappa = E(X_1) \vee 0 = \psi'(0+) \vee 0$.

Ultimately the factorization (34) can also be derived by a procedure of analytical extension and taking limits as q tends to zero in (18), simultaneously making use of the identities (20) and (21). A deeper look at this derivation yields the additional information that ϕ is the Laplace exponent of $\widehat{H}$; namely $\phi(\theta) = -\log E\left(e^{-\theta\widehat{H}_1}\right)$.

In the special case that $\Phi(0) = 0$, that is to say, the process X does not drift to $-\infty$ or equivalently that $\psi'(0+) \geq 0$, it can be shown that the scale function W describes the potential measure of $\widehat{H}$. Indeed, recall that the potential measure of $\widehat{H}$ is defined by

$$\int_0^\infty dt \cdot P(\widehat{H}_t \in dx), \qquad \text{for } x \geq 0. \tag{37}$$

Calculating its Laplace transform we get the identity

$$\int_0^\infty \int_0^\infty dt \cdot P(\widehat{H}_t \in dx) e^{-\theta x} = \frac{1}{\phi(\theta)} \qquad \text{for } \theta > 0. \tag{38}$$

Inverting the Laplace transform on the left hand side we get the identity

$$W(x) = \int_0^\infty dt \cdot P(\widehat{H}_t \in [0, x]), \qquad x \geq 0. \tag{39}$$

It can be shown similarly that in general, when $\Phi(0) \geq 0$, the scale function is related to the potential measure of $\widehat{H}$ by the formula

$$W(x) = e^{\Phi(0)x} \int_0^x e^{-\Phi(0)y} \int_0^\infty dt \cdot P(\widehat{H}_t \in dy), \qquad x \geq 0. \tag{40}$$

The connection between W and the potential measure of $\widehat{H}$ turns out to be of importance at several points later on in this exposition.

2.5 A Suite of Fluctuation Identities

Let us now show that scale functions are a natural family of functions with which one may describe a whole suite of fluctuation identities concerning first and last passage problems. We do not offer full proofs, concentrating instead on the basic ideas that drive the results. These are largely based upon the use of the Strong Markov Property and earlier established results on the law of $\overline{X}_{e_q}$ and $-\underline{X}_{e_q}$.

Many of the results in this section are taken from the recent work [9, 16, 82, 83]. However some of the identities can also be found in the setting of a compound Poisson jump structure from earlier Soviet–Ukranian work; see for example [25, 56–60, 98] to name but a few.

2.5.1 First Passage Problems

Recall that for all $a \in \mathbb{R}$, $\tau_a^+ = \inf\{t > 0 : X_t > a\}$ and $\tau_0^- = \inf\{t > 0 : X_t < 0\}$. We also introduce for each $x, q \geq 0$,

$$Z^{(q)}(x) = 1 + q \int_0^x W^{(q)}(y) dy. \tag{41}$$

Theorem 2.6 (One- and two-sided exit formulae).

(i) For any $x \in \mathbb{R}$ and $q \geq 0$,

$$E_x \left(e^{-q\tau_0^-} 1_{(\tau_0^- < \infty)} \right) = Z^{(q)}(x) - \frac{q}{\Phi(q)} W^{(q)}(x) , \tag{42}$$

where we understand $q/\Phi(q)$ in the limiting sense for $q = 0$, so that

$$P_x \left(\tau_0^- < \infty \right) = \begin{cases} 1 - \psi'(0+)W(x), & \text{if } \psi'(0+) > 0, \\ 1, & \text{if } \psi'(0+) \leq 0. \end{cases} \tag{43}$$

(ii) For all $x \geq 0$ and $q \geq 0$

$$E_x \left(e^{-q\tau_0^-} 1_{(X_{\tau_0^-} = 0)} \right) = \frac{\sigma^2}{2} \left\{ W^{(q)\prime}(x) - \Phi(q)W^{(q)}(x) \right\}, \tag{44}$$

where the right hand side is understood to be identically zero if $\sigma = 0$ and otherwise the derivative is well defined thanks to Lemma 2.4.
(iii) For any $x \leq a$ and $q \geq 0$,

$$E_x \left(e^{-q\tau_0^-} 1_{(\tau_0^- < \tau_a^+)} \right) = Z^{(q)}(x) - Z^{(q)}(a)\frac{W^{(q)}(x)}{W^{(q)}(a)}. \tag{45}$$

Sketch proof.

(i) Making use of Corollary 2.2 we have for $q > 0$ and $x \geq 0$,

$$E_x \left(e^{-q\tau_0^-} 1_{(\tau_0^- < \infty)} \right) = P_x \left(e_q > \tau_0^- \right)$$

$$= P_x \left(\underline{X}_{e_q} < 0 \right)$$

$$= P \left(-\underline{X}_{e_q} > x \right)$$

$$= 1 - P \left(-\underline{X}_{e_q} \leq x \right)$$

$$= 1 + q \int_0^x W^{(q)}(y)dy - \frac{q}{\Phi(q)} W^{(q)}(x)$$

$$= Z^{(q)}(x) - \frac{q}{\Phi(q)} W^{(q)}(x). \tag{46}$$

The proof for the case $q = 0$ follows by taking limits as $q \downarrow 0$ on both sides of the final equality in (46).

(ii) Suppose that $q = 0$ and $\psi'(0+) \geq 0$ (equivalently the descending ladder height process is not killed) then the claimed identity reads

$$P_x\left(X_{\tau_0^-} = 0\right) = \frac{\sigma^2}{2}W'(x), \qquad x \geq 0. \qquad (47)$$

Note that the probability on the left hand side is also equal to the probability that the descending ladder height subordinator creeps over x, $P(\widehat{H}_{\widehat{T}_x} = x)$, where $\widehat{T}_x = \inf\{t \geq 0 : \widehat{H}_t > x\}$. The identity (47) now follows directly from a classic result of Kesten [53] which shows that the probability that a subordinator creeps is non-zero if and only its drift coefficient is strictly positive, in which case it is equal to the drift coefficient multiplied by its potential density, which necessarily exists. This, together with (36) and (40), implies that $P(\widehat{H}_{\widehat{T}_x} = x) = \sigma^2 W'(x)/2$.

When $q > 0$ or $\psi'(0+) < 0$, the formula is proved using the change of measure (16) with $c = \Phi(q)$, $q \geq 0$, and the above result together with (25).

(iii) Fix $q > 0$. We have for $x \geq 0$,

$$E_x\left(e^{-q\tau_0^-}\mathbf{1}_{(\tau_0^- < \tau_a^+)}\right) = E_x\left(e^{-q\tau_0^-}\mathbf{1}_{(\tau_0^- < \infty)}\right) - E_x\left(e^{-q\tau_0^-}\mathbf{1}_{(\tau_a^+ < \tau_0^-)}\right).$$

Applying the Strong Markov Property at τ_a^+ and using the fact that X creeps upwards, we also have that

$$E_x\left(e^{-q\tau_0^-}\mathbf{1}_{(\tau_a^+ < \tau_0^-)}\right) = E_x\left(e^{-q\tau_a^+}\mathbf{1}_{(\tau_a^+ < \tau_0^-)}\right)E_a\left(e^{-q\tau_0^-}\mathbf{1}_{(\tau_0^- < \infty)}\right).$$

Piecing the previous two equalities together and appealing to (42) and (5) yields the desired conclusion. The case that $q = 0$ is again handled by taking limits as $q \downarrow 0$ on both sides of (45). Here we have used the discussion in Sect. 2.4.   □

Note that part (ii) above tallies with the earlier mentioned fact that spectrally negative Lévy processes do not creep downwards unless they have a Gaussian component. Note also that when $\sigma \neq 0$, Lemma 2.4 tells us that the derivative appearing on the right hand side of the density is everywhere defined for $x \geq 0$.

We also give expressions for the expected occupation measure of X in a given Borel set over its entire lifetime as well as when time is restricted up to the first passage times τ_a^+, τ_0^- and

$$\tau := \tau_a^+ \wedge \tau_0^-.$$

Such expected occupation measures are generally referred to as resolvents and play an important role in establishing, for example, deeper identities concerning first passage problems such as the one presented in Theorem 1.4.

Theorem 2.7 (Resolvents).

(i) *For all $a \geq x \geq 0$, $q \geq 0$ and Borel set $A \subseteq [0, a]$,*

$$E_x \left[\int_0^\infty e^{-qt} \mathbf{1}_{\{X_t \in A, \, t < \tau\}} dt \right]$$
$$= \int_A \left\{ \frac{W^{(q)}(x) W^{(q)}(a-y)}{W^{(q)}(a)} - W^{(q)}(x-y) \right\} dy.$$

(ii) *For all $a \geq x$ and Borel sets $A \subseteq (-\infty, a]$,*

$$E_x \left[\int_0^\infty e^{-qt} \mathbf{1}_{\{X_t \in A, \, t < \tau_a^+\}} dt \right]$$
$$= \int_A \left\{ e^{-\Phi(q)(a-x)} W^{(q)}(a-y) - W^{(q)}(x-y) \right\} dy.$$

(iii) *For all $x \geq 0$ and Borel set $A \subseteq [0, \infty)$,*

$$E_x \left[\int_0^\infty e^{-qt} \mathbf{1}_{\{X_t \in A, \, t < \tau_0^-\}} dt \right] = \int_A \left\{ e^{-\Phi(q)y} W^{(q)}(x) - W^{(q)}(x-y) \right\} dy.$$

(iv) *For all Borel set $A \subseteq \mathbb{R}$*

$$E \left[\int_0^\infty e^{-qt} \mathbf{1}_{\{X_t \in A\}} dt \right] = \int_A \left\{ (\psi'(\Phi(q)+))^{-1} e^{-\Phi(q)y} - W^{(q)}(-y) \right\} dy.$$

Sketch proof. We give an outline of the proof of (iii) from which the proof of (i) easily follows. The remaining two identities can be obtained by taking limits of the barriers in (i) relative to the initial position. Further details are given below. As usual we shall perform the relevant analysis in the case that $q > 0$. The case that $q = 0$ follows by taking limits as $q \downarrow 0$.

We start by noting that for all $x, y \geq 0$ and $q > 0$,

$$R^{(q)}(x, dy) := \int_0^\infty e^{-qt} P_x \left(X_t \in dy, \tau_0^- > t \right) dt = \frac{1}{q} P_x \left(X_{e_q} \in dy, \underline{X}_{e_q} \geq 0 \right),$$

where e_q is an independent, exponentially distributed random variable with parameter $q > 0$.

Appealing to the Wiener–Hopf factorization, specifically that $X_{e_q} - \underline{X}_{e_q}$ is independent of $\underline{X}_{e_q}$, and that $X_{e_q} - \underline{X}_{e_q}$ is equal in distribution to $\overline{X}_{e_q}$, we have that

$$R^{(q)}(x, dy) = \frac{1}{q} P((X_{\mathbf{e}_q} - \underline{X}_{\mathbf{e}_q}) + \underline{X}_{\mathbf{e}_q} \in dy - x, -\underline{X}_{\mathbf{e}_q} \leq x)$$

$$= \frac{1}{q} \int_{[0,x]} P\left(-\underline{X}_{\mathbf{e}_q} \in dz\right) P\left(\overline{X}_{\mathbf{e}_q} \in dy - x + z\right) 1_{\{y \geq x - z\}}.$$

Recall however, that $\overline{X}_{\mathbf{e}_q}$ is exponentially distributed with parameter $\Phi(q)$. In addition, the law of $-\underline{X}_{\mathbf{e}_q}$ has been identified in Corollary 2.2. Putting the pieces together and making some elementary manipulations the identity in (iii) follows.

Now suppose we denote the left hand side of the identity in (i) by $U^{(q)}(x, A)$. With the help of the Strong Markov Property we have that

$$qU^{(q)}(x, dy) = P_x\left(X_{\mathbf{e}_q} \in dy, \underline{X}_{\mathbf{e}_q} \geq 0, \overline{X}_{\mathbf{e}_q} \leq a\right)$$

$$= P_x\left(X_{\mathbf{e}_q} \in dy, \underline{X}_{\mathbf{e}_q} \geq 0\right)$$

$$-P_x\left(X_{\mathbf{e}_q} \in dy, \underline{X}_{\mathbf{e}_q} \geq 0, \overline{X}_{\mathbf{e}_q} > a\right)$$

$$= P_x\left(X_{\mathbf{e}_q} \in dy, \underline{X}_{\mathbf{e}_q} \geq 0\right)$$

$$-P_x\left(X_\tau = a, \tau < \mathbf{e}_q\right) P_a\left(X_{\mathbf{e}_q} \in dy, \underline{X}_{\mathbf{e}_q} \geq 0\right).$$

The first and third of the three probabilities on the right-hand side above have been computed in the previous paragraph, the second probability may be written

$$E_x\left(e^{-q\tau_a^+}; \tau_a^+ < \tau_0^-\right) = \frac{W^{(q)}(x)}{W^{(q)}(a)}.$$

The result now follows by assembling the relevant pieces.

To deduce (ii) from (i) we observe that if $a \geq x \geq 0$, $y > 0$ and $A \subset (-y, a]$ then (i) allows to determine the value of the expression

$$E_x\left[\int_0^\infty e^{-qt} 1_{\{X_t \in A, \, t < \tau_{-y}^- \wedge \tau_a^+\}} dt\right]$$

$$= E_{x+y}\left[\int_0^\infty e^{-qt} 1_{\{X_t \in A+y, \, t < \tau_0^- \wedge \tau_{a+y}^+\}} dt\right]$$

$$= \int_A \left\{\frac{W^{(q)}(x+y)}{W^{(q)}(a+y)} W^{(q)}(a-u) - W^{(q)}(x-u)\right\} du.$$

Then we use Lemma 3.3 to infer the following limit

$$\frac{W^{(q)}(x+y)}{W^{(q)}(a+y)} \xrightarrow[y \to \infty]{} e^{-\Phi(q)(a-x)}.$$

The result follows by a dominated convergence argument.

Finally (iv) follows from (ii) by taking $x = 0$, making a tend to infinity and using the estimate

$$\lim_{a \to \infty} e^{-\Phi(q)a} W^{(q)}(a - y) = \frac{e^{-\Phi(q)y}}{\psi'(\Phi(q))},$$

which is also a consequence of Lemma 3.3. $\qquad\square$

On a final note, the above resolvents easily lead to further identities of the type given in Theorem 1.4. Indeed, suppose that N is the Poisson random measure associated with the jumps of X. That is to say, N is a Poisson random measure on $[0, \infty) \times (-\infty, 0)$ with intensity $dt \times \Pi(dx)$. Recall that $\tau := \tau_a^+ \wedge \tau_0^-$. Then with the help of the Compensation Formula, we have that for $x \in [0, a]$, A any Borel set in $[0, a)$ and B any Borel set in $(-\infty, 0)$ and $q \geq 0$,

$$E_x \left(e^{-q\tau}; X_\tau \in B, X_{\tau-} \in A \right)$$

$$= E_x \left(\int_{[0,\infty)} \int_{(-\infty,0)} e^{-qt} \mathbf{1}_{(\overline{X}_{t-} \leq a, \underline{X}_{t-} \geq 0, X_{t-} \in A)} \mathbf{1}_{(y \in B - X_{t-})} N(dt \times dy) \right)$$

$$= E_x \left(\int_0^\infty e^{-qt} \mathbf{1}_{(t<\tau)} \Pi(B - X_t) \mathbf{1}_{(X_t \in A)} dt \right)$$

$$= \int_A \Pi(B - y) U^{(q)}(x, dy), \tag{48}$$

where, as noted earlier,

$$U^{(q)}(x, dy) = \int_0^\infty e^{-qt} P_x (X_t \in dy, \tau > t) \, dt.$$

Observe that the fact that B is a Borel subset of $(-\infty, 0)$, allow us not to consider the event where the process leaves the interval from below by creeping. Nevertheless, the probability of this event has been calculated in (44).

2.5.2 First Passage Problems for Reflected Processes

The list of fluctuation identities continues when one considers the reflected processes $\overline{Y} := \{\overline{X}_t - X_t : t \geq 0\}$ and $\underline{Y} := \{X_t - \underline{X}_t : t \geq 0\}$. Note that it is easy to prove that both of these processes are non-negative strong Markov processes. We shall henceforth denote their probabilities by $\{\overline{P}_x : x \in [0, \infty)\}$ and $\{\underline{P}_x : x \in [0, \infty)\}$. Note that $(\overline{Y}, \overline{P}_x)$ is equal in law to $\{(x \vee \overline{X}_t - X_t) : t \geq 0\}$ under P and $(\underline{Y}, \underline{P}_x)$ is equal in law to $\{X_t - (\underline{X}_t \wedge -x) : t \geq 0\}$ under P. The following theorem is a compilation of results taken from [9] and [82]. We do not offer proofs but instead we shall settle for remarking that the proofs use similar techniques to those of the previous section.

For convenience, let us define for $a > 0$,

$$\underline{\sigma}_a = \inf\{t \geq 0 : \underline{Y}_t > a\} \text{ and } \overline{\sigma}_a = \inf\{t \geq 0 : \overline{Y}_t > a\},$$

where $\underline{Y}_t = X_t - \underline{X}_t$ and $\overline{Y}_t = \overline{X}_t - X_t$.

Theorem 2.8. *Suppose that $a > 0$, $x \in [0, a]$ and $q \geq 0$. Then*

(i) $\underline{E}_x(e^{-q\underline{\sigma}_a}) = Z^{(q)}(x)/Z^{(q)}(a)$,

(ii) *Taking $W_+^{(q)\prime}(a)$ as the right derivative of $W^{(q)}$ at a,*

$$\overline{E}_x(e^{-q\overline{\sigma}_a}) = Z^{(q)}(a - x) - qW^{(q)}(a - x)W^{(q)}(a)/W_+^{(q)\prime}(a),$$

(iii) *For any Borel set $A \in [0, a)$,*

$$\underline{E}_x\left[\int_0^\infty e^{-qt} \mathbf{1}_{\{\underline{Y}_t \in A, \, t < \underline{\sigma}_a\}} dt\right]$$

$$= \int_A \left\{\frac{Z^{(q)}(x)}{Z^{(q)}(a)} W^{(q)}(a - y) - W^{(q)}(x - y)\right\} dy,$$

and

(iv) *For any Borel set $A \in [0, a)$,*

$$\overline{E}_x\left[\int_0^\infty e^{-qt} \mathbf{1}_{\{\overline{Y}_t \in A, \, t < \overline{\sigma}_a\}} dt\right]$$

$$= \int_A \left\{W^{(q)}(x - a)\frac{W_+^{(q)\prime}(y)}{W_+^{(q)\prime}(a)} - W^{(q)}(y - x)\right\} dy$$

$$+ \int_A \left\{W^{(q)}(x - a)\frac{W^{(q)}(0)}{W_+^{(q)\prime}(a)}\right\} \delta_0(dy).$$

3   Further Analytical Properties of Scale Functions

3.1   Behaviour at 0 and $+\infty$

Ultimately we are interested in describing the "shape" of scale functions. We start by looking at their behaviour at the origin and $+\infty$. In order to state the results more precisely, we recall from (15) that when X has paths of bounded variation, we may write it in the form $X_t = \delta t - S_t$, $t \geq 0$, where $\delta > 0$ and S is a pure jump subordinator.

Lemma 3.1. *For all $q \geq 0$, $W^{(q)}(0) = 0$ if and only if X has unbounded variation. Otherwise, when X has bounded variation, $W^{(q)}(0) = 1/\delta$.*

Proof. Note that for all $q > 0$,

$$
\begin{aligned}
W^{(q)}(0) &= \lim_{\beta \uparrow \infty} \int_0^\infty \beta \, e^{-\beta x} W^{(q)}(x) \mathrm{d}x \\
&= \lim_{\beta \uparrow \infty} \frac{\beta}{\psi(\beta) - q} \\
&= \lim_{\beta \uparrow \infty} \frac{\beta}{\psi(\beta)}.
\end{aligned}
\tag{49}
$$

Recall the spatial Wiener–Hopf factorization of ψ in (34),

$$
\psi(\beta) = (\beta - \Phi(0))\phi(\beta), \qquad \beta \geq 0,
$$

and that ϕ denotes the Laplace exponent of the downward ladder height subordinator $\widehat{H}$. It follows that

$$
W^{(q)}(0) = \lim_{\beta \uparrow \infty} \frac{\beta}{\psi(\beta)} = \lim_{\beta \to \infty} \frac{1}{\phi(\beta)}.
$$

Now, observe that $\lim_{\beta \to \infty} \phi(\beta) < \infty$, if and only if the Lévy measure of $\widehat{H}$ is a finite measure and its drift is 0. We know this happens if and only if X has paths of bounded variation. Indeed, this is the only case in which 0 is irregular for $(-\infty, 0)$ and hence starting from 0 it takes a strictly positive amount of time to enter the open lower half line. Accordingly there is finite activity over each finite time horizon for $\widehat{H}$. The first claim of the Theorem follows. Now, assume that X has paths of bounded variation. In this case, one may use (15) to write more simply

$$
\psi(\beta) = \delta\beta - \int_{(0,\infty)} (1 - e^{-\beta x}) \Pi(\mathrm{d}x).
$$

An integration by parts tells us that

$$
\frac{\psi(\beta)}{\beta} = \delta - \int_0^\infty e^{-\beta x} \Pi(-\infty, -x) \mathrm{d}x,
\tag{50}
$$

and hence, noting that bounded variation paths (equivalently $\sigma = 0$ and $\int_{(-1,0)} |x| \Pi(\mathrm{d}x) < \infty$) necessarily implies that $\int_{(0,1)} \Pi(-\infty, -x) \mathrm{d}x < \infty$, it follows that $\psi(\beta)/\beta \to \delta$ as $\beta \uparrow \infty$. From (49) we now see that, for all $q \geq 0$, $W^{(q)}(0) = 1/\delta$ as claimed.

$\square$

Returning to (29) we see that the conclusion of the previous lemma indicates that, precisely when X has bounded variation,

$$P_0\left(\tau_a^+ < \tau_0^-\right) = \frac{W(0)}{W(a)} > 0. \tag{51}$$

Note that the stopping time τ_0^- is defined with strict entry into $(-\infty, 0)$. Hence when X has the property that 0 is irregular for $(-\infty, 0)$, it takes an almost surely positive amount of time to exit the half line $[0, \infty)$. Since the aforementioned irregularity is equivalent to bounded variation for this class of Lévy processes, we see that (51) makes sense.

Next we turn to the behaviour of the right derivative of $W^{(q)}$, written $W_+^{(q)'}$, at 0.

Lemma 3.2. *For all $q \geq 0$ we have*

$$W_+^{(q)'}(0) = \begin{cases} 2/\sigma^2, & \text{when } \sigma \neq 0 \text{ or } \Pi(-\infty, 0) = \infty \\ (\Pi(-\infty, 0) + q)/\delta^2 & \text{when } \sigma = 0 \text{ and } \Pi(-\infty, 0) < \infty, \end{cases}$$

where we understand the first case to be $+\infty$ when $\sigma = 0$.

Proof. Using Lemma 2.3 and integrating (4) by parts we find that for $\theta > \Phi(q)$,

$$W^{(q)}(0+) + \int_0^\infty e^{-\theta x} W^{(q)'}(x) dx = \frac{\theta}{\psi(\theta) - q}. \tag{52}$$

Using this and a standard Tauberian theorem for Laplace transforms (see for instance [105] page 192, Theorem 4.3) we get

$$\begin{aligned} W_+^{(q)'}(0) &= \lim_{h \to 0+} \frac{W^{(q)}(h) - W^{(q)}(0)}{h} \\ &= \lim_{\theta \to \infty} \theta \int_0^\infty e^{-\theta x} W^{(q)'}(x) dx \\ &= \lim_{\theta \to \infty} \left(\frac{\theta^2}{\psi(\theta) - q} - \theta W^{(q)}(0+) \right). \end{aligned}$$

Now assume that X has paths of unbounded variation so that $W^{(q)}(0+) = 0$. We have for each $q \geq 0$,

$$W_+^{(q)'}(0) = \lim_{\theta \uparrow \infty} \int_0^\infty \theta e^{-\theta x} W^{(q)'}(x) dx = \lim_{\theta \uparrow \infty} \frac{\theta^2}{\psi(\theta) - q}.$$

Then dividing (34) by θ^2 it is easy to prove using (36) that the limit above is equal to $2/\sigma^2$ as required. The expression for $W^{(q)'}(0+)$ in the first case now follows.

When X has bounded variation, a little more care is needed. Recall that in this case $W^{(q)}(0+) = 1/\delta$, we have,

$$W_+^{(q)\prime}(0)$$

$$= \lim_{\beta\uparrow\infty} \frac{\beta^2}{\delta\beta - \beta \int_0^\infty e^{-\beta x} \Pi(-\infty, -x)\mathrm{d}x - q} - \beta W^{(q)}(0+)$$

$$= \lim_{\beta\uparrow\infty} \frac{\beta^2 \left(1 - W^{(q)}(0+)\delta + W^{(q)}(0+)\int_0^\infty e^{-\beta x} \Pi(-\infty, -x)\mathrm{d}x\right) + q\beta W^{(q)}(0+)}{\delta\beta - \int_0^\infty \beta e^{-\beta x} \Pi(-\infty, -x)\mathrm{d}x + q}$$

$$= \lim_{\beta\uparrow\infty} \frac{1}{\delta} \frac{\int_0^\infty \beta e^{-\beta x} \Pi(-\infty, -x)\mathrm{d}x + q}{\delta - \int_0^\infty e^{-\beta x} \Pi(-\infty, -x)\mathrm{d}x}$$

$$= \frac{\Pi(-\infty, 0) + q}{\delta^2}.$$

In particular, if $\Pi(-\infty, 0) = \infty$, then the right hand side above is equal to ∞ and otherwise, if $\Pi(-\infty, 0) < \infty$, then $W^{(q)\prime}(0+)$ is finite and equal to $(\Pi(-\infty, 0) + q)/\delta^2$. $\qquad\square$

Next we look at the asymptotic behaviour of the scale function at $+\infty$.

Lemma 3.3. *For $q \geq 0$ we have,*

$$\lim_{x\to\infty} e^{-\Phi(q)x} W^{(q)}(x) = 1/\psi'(\Phi(q)),$$

and

$$\lim_{x\to\infty} Z^{(q)}(x)/W^{(q)}(x) = q/\Phi(q),$$

where the right hand side above is understood in the limiting sense $\lim_{q\downarrow 0} q/\Phi(q) = 0 \vee (1/\psi'(0+))$ when $q = 0$.

Proof. For the first part, recall the identity (33) which is valid for all $q \geq 0$. It follows from (22) that

$$W^{(q)}(x) = e^{\Phi(q)x} \frac{1}{\psi'_{\Phi(q)}(0+)} P_x^{\Phi(q)} \left(\underline{X}_\infty \geq 0\right). \tag{53}$$

Appealing to (26) we note that $\psi'_{\Phi(q)}(0+) = \psi'(\Phi(q)) > 0$ (which in particular implies that X under $\mathbb{P}^{\Phi(q)}$ drifts to $+\infty$) and hence

$$\lim_{x\uparrow\infty} e^{-\Phi(q)x} W^{(q)}(x) = \frac{1}{\psi'(\Phi(q))} \lim_{x\uparrow\infty} P_x^{\Phi(q)} \left(\underline{X}_\infty \geq 0\right) = \frac{1}{\psi'(\Phi(q))}.$$

Note that from this proof we also see that $W_{\Phi(q)}(+\infty) = 1/\psi'(\Phi(q))$.

For the second part, it suffices to compare the identity (45) as $a \uparrow \infty$ against (42).   □

3.2   Concave–Convex Properties

Lemma 3.3 implies that when $q > 0$, $W^{(q)}$ grows in a way that is asymptotically exponential and this opens the question as to whether there are any convexity properties associated with such scale functions for large values. Numerous applications have shown the need to specify more detail about the shape, and ultimately, the smoothness of scale functions. See for example the discussion in Chap. 1. In this respect, there has been a recent string of articles, each one improving of the last, which have investigated concavity–convexity properties of scale functions and which are based on the following fundamental observation of Renming Song. Suppose that $\psi'(0+) \geq 0$ and that

$$\overline{\Pi}(x) := \Pi(-\infty, -x),$$

the density of the Lévy measure of the descending ladder height process, see (36), is completely monotone. Amongst other things, this implies that the descending ladder height process $\widehat{H}$ belongs to the class of so-called complete subordinators. A convenience of this class of subordinators is that the potential measure associated to $\widehat{H}$, which in this case is W, has a derivative which is completely monotone; see for example Song and Vondraček [95]. In particular this implies that W is concave (as well as being infinitely smooth). Loeffen [75] pushes this idea further to the case of $W^{(q)}$ for $q > 0$ as follows. (Similar ideas can also be developed from the paper of Rogers [87]).

Theorem 3.4. *Suppose that $-\overline{\Pi}$ has a density which is completely monotone then $W^{(q)}$ has a density on $(0, \infty)$ which is strictly convex. In particular, this implies the existence of a constant a^* such that $W^{(q)}$ is strictly concave on $(0, a^*)$ and strictly convex on (a^*, ∞).*

Proof. Recall from (25) that we may always write $W^{(q)}(x) = e^{\Phi(q)x} W_{\Phi(q)}(x)$, where $W_{\Phi(q)}$ plays the role of W under the measure $P^{\Phi(q)}$. Recall also from the discussion under (26) and (17) that $(X, P^{\Phi(q)})$ drifts to $+\infty$, and that the Lévy measure of X under $P^{\Phi(q)}$ is given by $\Pi_{\Phi(q)}(\mathrm{d}x) = e^{\Phi(q)x}\Pi(\mathrm{d}x)$, $x \in \mathbb{R}$. It follows that under the assumptions of the Theorem that the function $\Pi_{\Phi(q)}(-\infty, -x)$, $x > 0$, is completely monotone. Hence the discussion preceding the statement of Theorem 3.4 tells us that $W'_{\Phi(q)}(x)$ exists and is completely monotone. According to the definition by Song and Vondraček [95], this makes of $W_{\Phi(q)}$ a *Bernstein function*.

The general theory of Bernstein functions dictates that there necessarily exists a triple (a, b, ξ), where $a, b \geq 0$ and ξ is a measure concentrated on $(0, \infty)$ satisfying

$\int_{(0,\infty)} (1 \wedge t) \xi(\mathrm{d}t) < \infty$, such that

$$W_{\Phi(q)}(x) = \mathrm{a} + \mathrm{b}x + \int_{(0,\infty)} (1 - \mathrm{e}^{-xt}) \xi(\mathrm{d}t).$$

It is now a straightforward exercise to check with the help of the above identity and the Dominated Convergence Theorem that for $x > 0$,

$$W^{(q)\prime\prime\prime}(x) = f'''(x) + \int_{(0,\Phi(q)]} (\Phi(q)^3 \mathrm{e}^{\Phi(q)x} - (\Phi(q) - t)^3 \mathrm{e}^{(\Phi(q)-t)x}) \xi(\mathrm{d}t)$$

$$+ \int_{(\Phi(q),\infty)} (\Phi(q)^3 \mathrm{e}^{\Phi(q)x} + (t - \Phi(q))^3 \mathrm{e}^{-x(t-\Phi(q))}) \xi(\mathrm{d}t),$$

where $f(x) = (\mathrm{a} + \mathrm{b}x)\mathrm{e}^{\Phi(q)x}$. Hence $W^{(q)\prime\prime\prime}(x) > 0$ for all $x > 0$, showing that $W^{(q)\prime}$ is strictly convex on $(0, \infty)$ as required. $\qquad\square$

Following additional contributions in [70], the final word on concavity–convexity currently stands with the following theorem, taken from [78], which overlaps all of the aforementioned results.

Theorem 3.5. *Suppose that $\overline{\Pi}$ is log-convex. Then for all $q \geq 0$, $W^{(q)}$ has a log-convex first derivative.*

Note that the existence of a log-convex density of $-\overline{\Pi}$ implies that $\overline{\Pi}$ is log-convex and hence the latter is a weaker condition than the former. This is not an obvious statement, but it can be proved using elementary analytical arguments.

3.3 Analyticity in q

Let us now look at the behaviour of $W^{(q)}$ as a function in q.

Lemma 3.6. *For each $x \geq 0$, the function $q \mapsto W^{(q)}(x)$ may be analytically extended to $q \in \mathbb{C}$.*

Proof. For a fixed choice of $q > 0$,

$$\int_0^\infty \mathrm{e}^{-\beta x} W^{(q)}(x)\mathrm{d}x = \frac{1}{\psi(\beta) - q}$$

$$= \frac{1}{\psi(\beta)} \frac{1}{1 - q/\psi(\beta)}$$

$$= \frac{1}{\psi(\beta)} \sum_{k \geq 0} q^k \frac{1}{\psi(\beta)^k}, \qquad (54)$$

for $\beta > \Phi(q)$. The latter inequality implies that $0 < q/\psi(\beta) < 1$. Next note that

$$\sum_{k \geq 0} q^k W^{*(k+1)}(x)$$

converges for each $x \geq 0$ where W^{*k} is the kth convolution of W with itself. This is easily deduced once one has the estimates

$$W^{*(k+1)}(x) \leq \frac{x^k}{k!} W(x)^{k+1}, \qquad x \geq 0, \tag{55}$$

which can easily be established by induction. Indeed note that if (55) holds for $k \geq 0$, then by monotonicity of W,

$$\begin{aligned} W^{*(k+1)}(x) &\leq \int_0^x \frac{y^{k-1}}{(k-1)!} W(y)^k W(x-y) \, dy \\ &\leq \frac{1}{(k-1)!} W(x)^{k+1} \int_0^x y^{k-1} \, dy \\ &= \frac{x^k}{k!} W(x)^{k+1} . \end{aligned}$$

Returning to (54) we may now apply Fubini's Theorem (justified by the assumption that $\beta > \Phi(q)$) and deduce that

$$\begin{aligned} \int_0^\infty e^{-\beta x} W^{(q)}(x) dx &= \sum_{k \geq 0} q^k \frac{1}{\psi(\beta)^{k+1}} \\ &= \sum_{k \geq 0} q^k \int_0^\infty e^{-\beta x} W^{*(k+1)}(x) \, dx \\ &= \int_0^\infty e^{-\beta x} \sum_{k \geq 0} q^k W^{*(k+1)}(x) \, dx. \end{aligned}$$

Thanks to continuity of W and $W^{(q)}$ we have that

$$W^{(q)}(x) = \sum_{k \geq 0} q^k W^{*(k+1)}(x), \quad x \in \mathbb{R}. \tag{56}$$

Now noting that $\sum_{k \geq 0} q^k W^{*(k+1)}(x)$ converges for all $q \in \mathbb{C}$ we may extend the definition of $W^{(q)}$ for each fixed $x \geq 0$, by the equality given in (56). $\square$

For each $c \geq 0$, denote by $W_c^{(q)}$ the q-scale function for (X, P^c). The previous Lemma allows us to establish the following relationship for $W_c^{(q)}$ with different values of q and c.

Lemma 3.7. *For any $q \in \mathbb{C}$ and $c \in \mathbb{R}$ such that $|\psi(c)| < \infty$ we have*

$$W^{(q)}(x) = e^{cx}W_c^{(q-\psi(c))}(x), \tag{57}$$

for all $x \geq 0$.

Proof. For a given $c \in \mathbb{R}$, such that $|\psi(c)| < \infty$, the identity (57) holds for $q - \psi(c) \geq 0$, on account of both left and right-hand side being continuous functions with the same Laplace transform. By Lemma 3.6 both left- and right-hand side of (57) are analytic in q for each fixed $x \geq 0$. The Identity Theorem for analytic functions thus implies that they are equal for all $q \in \mathbb{C}$. □

Unfortunately a convenient relation such as (57) cannot be given for $Z^{(q)}$. Nonetheless we do have the following obvious corollary.

Corollary 3.8. *For each $x > 0$ the function $q \mapsto Z^{(q)}(x)$ may be analytically extended to $q \in \mathbb{C}$.*

The above results allow one to push some of the identities in Sect. 2.5 further by applying an exponential change of measure. In principle this allows one to gain distributional information about the position of the Lévy process at first passage. We give one example here but the reader can easily explore other possibilities.

Consider the first passage identity in (42). Suppose that $v \geq 0$, then $|\psi(v)| < \infty$, and assume it satisfies $u > \psi(v) \vee 0$. Then with the help of the aforementioned identity together with the change of measure (16) we have for $x \in \mathbb{R}$,

$$E_x\left(e^{-u\tau_0^- + vX_{\tau_0^-}}\mathbf{1}_{(\tau_0^- < \infty)}\right) = E_x^v\left(e^{-(u-\psi(v))\tau_0^-}\mathbf{1}_{(\tau_0^- < \infty)}\right)$$

$$= e^{vx}\left(Z_v^{(p)}(x) - \frac{p}{\Phi_v(p)}W_v^{(p)}(x)\right),$$

where $p = u - \psi(v) > 0$. We can develop the right hand side further using the relationship between scale functions given in Lemma 3.7 as well as by noting that

$$\Phi_v(p) = \sup\{\lambda \geq 0 : \psi_v(\lambda) = p\}$$
$$= \sup\{\lambda \geq 0 : \psi(\lambda + v) - \psi(v) = u - \psi(v)\}$$
$$= \sup\{\theta \geq 0 : \psi(\theta) = u\} - v$$
$$= \Phi(u) - v.$$

Hence we have that

$$E_x\left(e^{-u\tau_0^- + vX_{\tau_0^-}}\mathbf{1}_{(\tau_0^- < \infty)}\right)$$

$$= e^{vx}\left(1 + (u - \psi(v))\int_0^x e^{-vy}W^{(u)}(y)dy - \frac{u - \psi(v)}{\Phi(u) - v}e^{-vx}W^{(u)}(x)\right). \tag{58}$$

Clearly the restriction that $u > \psi(v)$ is an unnecessary constraint for both the left and right hand side of the above equality to be finite and it would suffice that u, $v \geq 0$. In particular for the right hand side, when $u = \psi(v)$ it follows that $\Phi(u) = v$ and hence the ratio $(u - \psi(v))/(\Phi(u) - v)$ should be understood in the limiting sense. That is to say,

$$\lim_{p \to 0} \frac{p}{\Phi_v(p)} = \lim_{p \to 0} \frac{\psi_v(\Phi_v(p))}{\Phi_v(p)} = \psi_v'(0+) = \psi'(v),$$

where we have used the fact that $\Phi_v(0) = 0$ as $\psi_v'(0+) = \psi'(v) > 0$.

Note that the left hand side of (58) is analytic for complex u with a strictly positive real part. Moreover, thanks to Lemma 3.6 and the fact that $\Phi(u)$ is a Laplace exponent (cf. the last equality of (19)) it is also clear that the right hand side can be analytically extended to allow for complex-valued u with strictly positive real part. Hence once again the Identity Theorem allows us to extend the equality (58) to allow for the case that $u > 0$. The case that $u = 0$ can be established by taking limits as $u \downarrow 0$ on both sides.

Careful inspection of the above argument shows that one may even relax the constraint on $v \geq 0$ to simply any v such that $|\psi(v)| < \infty$ in the case that the exponent $\psi(v)$ is finite for negative values of v.

3.4 Spectral Gap

Bertoin [17] showed an important consequence of the analytic nature of scale function $W^{(q)}$ in its argument q.

For $a > 0$, let $\tau := \tau_a^+ \wedge \tau_0^-$, then [17] investigates ergodicity properties of the spectrally negative Lévy process X killed on exiting $[0, a]$. The main object of concern is the killed transition kernel

$$P(x, t, A) = P_x\left(X_t \in A; t < \tau\right).$$

Theorem 3.9. *Define*

$$\rho = \inf\left\{q \geq 0 : W^{(-q)}(a) = 0\right\}.$$

Then ρ is finite and positive, and for any $q < \rho$ and $x \in (0, a)$, $W^{(-q)}(x) > 0$. Furthermore, the following assertions hold,

(i) *ρ is a simple root of the entire function $q \mapsto W^{(-q)}(a)$,*
(ii) *The function $W^{(-\rho)}$ is positive on $(0, a)$ and is ρ-invariant for $P(x, t, \cdot)$ in the sense that*

$$\int_{[0,a]} P(x, t, dy) W^{(-\rho)}(y) = e^{-\rho t} W^{(-\rho)}(x), \text{ for any } x \in (0, a),$$

(iii) *The measure $W^{(-\rho)}(a-x)\mathrm{d}x$ on $[0,a]$ is ρ-invariant in the sense that*

$$\int_{[0,a]} \mathrm{d}y \cdot W^{(-\rho)}(a-y)P(y,t,\mathrm{d}x) = e^{-\rho t}W^{(-\rho)}(a-x)\mathrm{d}x,$$

(iv) *There is a constant $c > 0$ such that, for any $x \in (0,a)$,*

$$\lim_{t\uparrow\infty} e^{\rho t}P(x,t,\mathrm{d}y) = cW^{(-\rho)}(x)W^{(-\rho)}(a-y)\mathrm{d}y,$$

in the sense of weak convergence.

A particular consequence of the part (iv) of the previous theorem is that there exists a constant $c' > 0$ such that

$$P_x\,(\tau > t) \sim c'W^{(-\rho)}(x)e^{-\rho t}.$$

as $t \uparrow \infty$. The constant ρ thus describes the rate of decay of the exit probability. By analogy with the theory of diffusions confined to compact domains, $-\rho$ also plays the role of the leading eigen-value, or spectral gap, of the infinitesimal generator of X constrained to the interval $(0,a)$ with Dirichlet boundary conditions. From part (i) of the above theorem, we see that the associated eigen-function is $W^{(-\rho)}$.

Lambert [72] strengthens this analogy with diffusions and showed further that for each $x \in (0,a)$,

$$e^{\rho t}\frac{W^{(-\rho)}(X_t)}{W^{(-\rho)}(x)}\mathbf{1}_{\{\tau>t\}}, \; t \geq 0$$

is a P_x martingale which, when used as a Radon–Nikodim density to change measure, induces a new probability measure, say, $P_x^\uparrow$. He shows moreover that this measure corresponds to the law of X conditioned to remain in $(0,a)$ in the sense that for all $t \geq 0$,

$$P_x^\uparrow(A) = \lim_{s\uparrow\infty} P_x(A|\tau > t + s), \;\; A \in \mathscr{F}_t.$$

The role of the scale function is no less important in other types of related conditioning. We have already alluded to its relevance to conditioning spectrally negative Lévy processes to stay positive in the first chapter. Pistorius [81, 82] also shows that a similar agenda to the above can be carried out with regard to reflected spectrally negative Lévy processes.

3.5   General Smoothness and Doney's Conjecture

Let $a > 0$, and recall $\tau = \tau_a^+ \wedge \tau_0^-$. It is not difficult to show from the identity (5) that for all $q \geq 0$,

$$e^{-q(t\wedge\tau)}W^{(q)}(X_{t\wedge\tau}), \; t \geq 0$$

is a martingale. Indeed, we have from the Strong Markov Property that for all $x \in \mathbb{R}$ and $t \geq 0$

$$E_x \left(e^{-q\tau_a^+} \mathbf{1}_{(\tau_0^- > \tau_a^+)} | \mathscr{F}_{t \wedge \tau_0^- \wedge \tau_a^+} \right)$$

$$= e^{-qt} \mathbf{1}_{\{t < \tau\}} E_{X_t} \left(e^{-q\tau_a^+} \mathbf{1}_{(\tau_0^- > \tau_a^+)} \right) + e^{-q\tau_a^+} \mathbf{1}_{\{\tau_a^+ < t \wedge \tau_0^-\}}$$

$$= e^{-q(t \wedge \tau_0^- \wedge \tau_a^+)} \frac{W^{(q)}(X_{t \wedge \tau_0^- \wedge \tau_a^+})}{W^{(q)}(a)},$$

where it should be noted that for the second equality we have used the fact that $W^{(q)}(X_{\tau_a^+})/W^{(q)}(a) = 1$ and $W^{(q)}(X_{\tau_0^-})/W^{(q)}(a) = 0$. Note in particular, for the last equality, X creeps downwards if and only if it has a Gaussian component in which case $W^{(q)}(X_{\tau_0^-}) = W^{(q)}(0+) = 0$, on the event of creeping, and otherwise on the event that $X_{\tau_0^-} < 0$, it trivially holds that $W^{(q)}(X_{\tau_0^-}) = 0$.

Naively speaking this reiterates an idea seen in the previous section that $W^{(q)}$ is an eigen-function with respect to the infinitesimal generator of X with eigenvalue q. That is to say, in an appropriate sense, $W^{(q)}$ solves the integro-differential equation

$$(\Gamma - q)W^{(q)}(x) = 0 \text{ on } (0, a),$$

where Γ is the infinitesimal generator of X, which is known to be given by

$$\Gamma f(x) = \mu f'(x) + \frac{1}{2}\sigma^2 f''(x)$$

$$+ \int_{(-\infty,0)} \left\{ f(x + y) - f(x) - f'(x)y\mathbf{1}_{\{y > -1\}} \right\} \Pi(dy),$$

for all f in its domain.

The problem with the above heuristic observation is that $W^{(q)}$ may not be in the domain of Γ. In particular, for the last equality to have a classical meaning we need at least that $f \in C^2(0, a)$ and $\int_{(-\infty,0)} f(x + y)\Pi(dy) < \infty$. This motivates the question of how smooth scale functions are. Given the dependency of concavity–convexity properties on the Lévy measure discussed in Sect. 3.2 as well as the statement of Lemma 2.4, it would seem sensible to believe that a relationship exists between the smoothness of the Lévy measure and the smoothness of the scale function $W^{(q)}$. In addition, one would also expect the inclusion of a Gaussian coefficient to have some effect on the smoothness of the scale function. Below we give a string of recent results which have attempted to address this matter. For notational convenience we shall write $W^{(q)} \in C^k(0, \infty)$ to mean that the restriction of $W^{(q)}$ to $(0, \infty)$ belongs to the $C^k(0, \infty)$ class.

Theorem 3.10. *Suppose that $\sigma^2 > 0$, then for all $q \geq 0$, $W^{(q)} \in C^2(0, \infty)$.*

Proof. As before, it is enough to consider the case where $q = 0$ and $\psi'(0+) \geq 0$. Indeed, in the case $q > 0$ or $q = 0$ and $\psi'(0+) < 0$, the identity (25) implies $W^{(q)}(x) = e^{\Phi(q)x}W_{\Phi(q)}(x)$, $x \in \mathbb{R}$; this together with the fact that $W_{\Phi(q)}$ is the 0 scale function of a spectrally negative Lévy process whose Laplace exponent, which is given by (26), satisfies the latter assumptions, allows us to easily deduce the result in these cases. We know from Lemma 2.4 that when $\sigma^2 > 0$, $W^{(q)} \in C^1(0, \infty)$. Recall from (32) that $W'(x) = n(\bar{\epsilon} \geq x)W(x)$ and hence if the limits exist, then

$$
\begin{aligned}
W''_+(x) &:= \lim_{\varepsilon \downarrow 0} \frac{W'(x+\varepsilon) - W'(x)}{\varepsilon} \\
&= -\lim_{\varepsilon \downarrow 0} \frac{n(\bar{\epsilon} \in [x, x+\varepsilon))\, W(x) - n(\bar{\epsilon} \geq x+\varepsilon)\,(W(x+\varepsilon) - W(x))}{\varepsilon} \\
&= -\lim_{\varepsilon \downarrow 0} \frac{n(\bar{\epsilon} \in [x, x+\varepsilon))}{\varepsilon} W(x) + n(\bar{\epsilon} \geq x)\, W'(x+). \quad (59)
\end{aligned}
$$

To show that the limit on the right hand side of (59) exists, define $\sigma_x = \inf\{t > 0 : \epsilon_t \geq x\}$ and $\mathscr{G}_t = \sigma(\epsilon_s : s \leq t)$. With the help of the Strong Markov Property for the excursion process we may write

$$
\begin{aligned}
n(\bar{\epsilon} \in [x, x+\varepsilon)) &= n(\sigma_x < \infty, \epsilon(\sigma_x) < x+\varepsilon, \bar{\epsilon} < x+\varepsilon) \\
&= n\left(1_{\{\sigma_x < \infty, \epsilon(\sigma_x) < x+\varepsilon\}} n(\bar{\epsilon} < x+\varepsilon | \mathscr{G}_{\sigma_x})\right) \\
&= n\left(1_{\{\sigma_x < \infty, \epsilon(\sigma_x) < x+\varepsilon\}} P_{-\epsilon(\sigma_x)}\left(\tau_0^+ < \tau_{-(x+\varepsilon)}^-\right)\right) \\
&= n(\sigma_x < \infty, \epsilon(\sigma_x) = x)\frac{W(\varepsilon)}{W(x+\varepsilon)} \\
&\quad + n\left(1_{\{\sigma_x < \infty, x < \epsilon(\sigma_x) < x+\varepsilon\}}\frac{W(x+\varepsilon - \epsilon(\sigma_x))}{W(x+\varepsilon)}\right). \quad (60)
\end{aligned}
$$

We know that spectrally negative Lévy processes which have a Gaussian component can creep downwards. Hence for the case at hand it follows that the event $\{\epsilon(\sigma_x) = x\}$ has non-zero n-measure.

Using the facts that $W'_+(0) = 2/\sigma^2$ and $W(0+) = 0$ (cf. Lemmas 3.1 and 3.2) together with the monotonicity of W, we have that

$$
\begin{aligned}
&\limsup_{\varepsilon \downarrow 0} \frac{1}{\varepsilon} n\left(1_{(\bar{\epsilon} \geq x, \epsilon(\sigma_x) \in (x, x+\varepsilon))}\frac{W(x+\varepsilon - \epsilon(\sigma_x))}{W(x+\varepsilon)}\right) \\
&\leq \frac{1}{W(x)} \limsup_{\varepsilon \downarrow 0} n(\bar{\epsilon} \geq x, \epsilon(\sigma_x) \in (x, x+\varepsilon))\frac{W(\varepsilon)}{\varepsilon} \quad (61) \\
&= 0.
\end{aligned}
$$

In conclusion, $W_+''(x)$ exists and

$$W_+''(x) = -W_+'(0)n\left(\bar{\epsilon} \ge x, \epsilon(\sigma_x) = x\right) + n\left(\bar{\epsilon} \ge x\right)W'(x),$$

that is to say,

$$n\left(\bar{\epsilon} \ge x, \epsilon(\sigma_x) = x\right) = \frac{\sigma^2}{2}\left\{\frac{W'(x)^2}{W(x)} - W_+''(x)\right\}. \tag{62}$$

We shall now show that there exists a left second derivative $W_-''(x)$ which may replace the role of $W_+''(x)$ on the right hand side of (62). For this we need to recall a description of the excursion measure as limit (see for instance Corollary 12 in page 88 in [36]) that states that for $A \in \mathscr{F}_t$

$$n\left(A, t < \zeta\right) = \lim_{y \downarrow 0} \frac{\widehat{P}_y(A, t < \tau_0^-)}{y},$$

and moreover this identity may be extended to stopping times. From this we may write

$$n\left(\bar{\epsilon} \ge x, \epsilon(\sigma_x) = x\right) = \lim_{y \downarrow 0} \frac{\widehat{P}_y\left(X_{\tau_x^+} = x; \tau_x^+ < \tau_0^-\right)}{y}, \tag{63}$$

where $\widehat{P}_y$ is the law of $-X$ when issued from y. From part (ii) of Theorem 2.6 we also know that

$$\widehat{P}_y\left(X_{\tau_x^+} = x; \tau_x^+ < \infty\right) = \frac{\sigma^2}{2}W'(x - y).$$

Using the Strong Markov Property, the above formula and the fact that X creeps upwards it is straightforward to deduce that

$$\widehat{P}_y\left(X_{\tau_x^+} = x; \tau_0^- < \tau_x^+\right) = \frac{W(x - y)}{W(x)} \times \frac{\sigma^2}{2}W'(x)$$

and hence

$$\widehat{P}_y(X_{\tau_x^+} = x; \tau_x^+ < \tau_0^-) = \frac{\sigma^2}{2}\left\{W'(x - y) - \frac{W(x - y)}{W(x)}W'(x)\right\}.$$

Returning to (63) we compute,

$$n\left(\bar{\epsilon} \ge x, \epsilon(\sigma_x) = x\right)$$

$$= \lim_{y \downarrow 0}\frac{\sigma^2}{2}\left\{\frac{W'(x)}{W(x)}\frac{W(x) - W(x - y)}{y} - \frac{W'(x) - W'(x - y)}{y}\right\}$$

$$= \frac{\sigma^2}{2}\left\{\frac{W'(x)^2}{W(x)} - W_-''(x)\right\}.$$

Note that the existence of $W''_-(x)$ is guaranteed in light of (63). We see that the final equality above is identical to (62) but with $W''_+(x)$ replaced by $W''_-(x)$.

Thus far we have shown that a second derivative exists everywhere. To complete the proof, we need to show that this second derivative is continuous. To do this, it suffices to show that $n\,(\bar{\epsilon} \geq x, \epsilon(\sigma_x) = x)$ is continuous. To this end note that a straightforward computation, similar to (63) but making use of Part (i) of Theorem 2.7 and L'Hôpital's rule to compute the relevant limit, shows that

$$n\,(\bar{\epsilon} \geq x, \epsilon(\sigma_x) > x) = \int_0^x \left\{ W'(x - y) - \frac{W'(x)}{W(x)} W(x - y) \right\} \overline{\varPi}(y) \mathrm{d}y,$$

where we recall that $\overline{\varPi}(y) = \varPi(-\infty, -y)$. Hence it is now easy to see that

$$n\,(\bar{\epsilon} \geq x, \epsilon(\sigma_x) = x) = n\,(\bar{\epsilon} \geq x) - n\,(\bar{\epsilon} \geq x, \epsilon(\sigma_x) > x)$$

is continuous, thus completing the proof. $\qquad\qquad\qquad\qquad\qquad\qquad\square$

Chan et al. [27] take a more analytical approach to the smoothness of scale functions by exploring its connection with the classical renewal equation. They note that, on account of (25), it suffices to consider smoothness in the case that $\psi'(0+) \geq 0$ (i.e. $\varPhi(0) = 0$). Within this regime, the basic idea is to start with the obvious statement that under the measure P_x the Lévy process will cross downwards below zero by either creeping across it or jumping clear below it. Taking account of (43), (44) and (48) in its limiting form as $a \uparrow \infty$, we thus have

$$1 - \frac{1}{\psi'(0+)} W(x) = P_x \left(\tau_0^- < \infty \right)$$

$$= P_x \left(X_{\tau_0^-} = 0 \right) + P_x \left(X_{\tau_0^-} < 0 \right)$$

$$= \frac{\sigma^2}{2} W'(x) + \int_0^\infty \{ W(x) - W(x - y) \} \overline{\varPi}(y) \mathrm{d}y$$

$$= \frac{\sigma^2}{2} W'(x) + \int_0^x W'(z) \overline{\overline{\varPi}}(x - z) \mathrm{d}z$$

$$= \frac{\sigma^2}{2} W'(x) + \int_0^x W'(x - z) \overline{\overline{\varPi}}(z) \mathrm{d}z, \qquad (64)$$

where $\overline{\overline{\varPi}}(z) = \int_z^\infty \overline{\varPi}(y) \mathrm{d}y$. Hence,

$$1 = \frac{\sigma^2}{2} W'(x) + \int_0^x W'(x - z) \left\{ \overline{\overline{\varPi}}(z) + \frac{1}{\psi'(0+)} \right\} \mathrm{d}z, \qquad (65)$$

and after a little manipulation, we arrive at the classical renewal equation

$$f = 1 + f \star g,$$

where $f(x) = \sigma^2 W'(x)/2$, and $g(x) = -2\sigma^{-2} \int_0^x \overline{\overline{\varPi}}(y)\mathrm{d}y - 2\sigma^{-2}\psi'(0+)x$, and momentarily we have assumed that $\sigma^2 > 0$. By engaging with the well known convolution series solution to the renewal equation they establish the following result, which goes beyond the scope of Theorem 3.10.

Theorem 3.11. *Suppose that X has a Gaussian component and its Blumenthal-Getoor index belongs to $[0, 2)$, that is to say*

$$\inf \left\{ \beta \geq 0 : \int_{|x|<1} |x|^\beta \varPi(\mathrm{d}x) < \infty \right\} \in [0, 2).$$

Then for each $q \geq 0$ and $n = 0, 1, 2, \ldots$, $W^{(q)} \in C^{n+3}(0, \infty)$ if and only if $\overline{\varPi} \in C^n(0, \infty)$.

In fact the method used to establish the above theorem can also be used to prove Theorem 3.10. It is also apparent from the analysis of [27] that their method cannot be applied when considering renewal equations of the form $1 = f \star g$, which would be the form of the resulting renewal equation in (65) when $\sigma = 0$ with an appropriate choice of f and g.

However when X does have paths of bounded variation, the aforementioned is possible following an integration by parts and a little algebra in the convolution on the right hand side of (64). In that case we take $f = \delta W(x)$ and $g(x) = \delta^{-1} \int_0^x \overline{\overline{\varPi}}(y)\mathrm{d}y$ where we recall that $W(0+) = \delta^{-1}$ and δ is the coefficient of the linear drift when X is written according to the decomposition (15). Note that this integration by parts is not possible if X has paths of unbounded variation with no Gaussian component as the aforementioned choice of g is not finite.

The same analysis for the Gaussian case but now for the bounded variation setting in [27] yields the following result.

Theorem 3.12. *Suppose that X has paths of bounded variation and $-\overline{\varPi}$ has a density $\pi(x)$, such that $\pi(x) \leq C|x|^{-1-\alpha}$ in the neighbourhood of the origin, for some $\alpha < 1$ and $C > 0$. Then for each $q \geq 0$ and $n = 1, 2, \ldots$, $W^{(q)} \in C^{n+1}(0, \infty)$ if and only if $\overline{\varPi} \in C^n(0, \infty)$.*

Essentially the results of [27] are primarily results about the smoothness of the solutions to renewal (or indeed Volterra) equations which are then applied where possible to the particular setting of scale functions. Unfortunately this means that nothing has been said about the case that X has paths of unbounded variation and $\sigma = 0$ to date. Moreover, the results in the last two theorems above are subject to conditions which do not appear to be probabilistically natural other than to provide the technical basis with which to push through their respective analytical proofs. Nonetheless they go part way to addressing *Doney's conjecture for scale functions*[1] as follows.

---

[1]Curious about the results in [27], Doney produced a number of specific examples of Lévy processes whose scale functions exhibited analytical behaviour that lead to his conjecture (personal communication with A.E. Kyprianou).

Conjecture 3.13. For $k = 0, 1, 2, \cdots$

1. If $\sigma^2 > 0$ then
$$W \in C^{k+3}(0, \infty) \Leftrightarrow \overline{\varPi} \in C^k(0, \infty),$$

2. If $\sigma = 0$ and $\int_{(-1,0)} |x| \varPi(\mathrm{d}x) = \infty$ then
$$W \in C^{k+2}(0, \infty) \Leftrightarrow \overline{\varPi} \in C^k(0, \infty),$$

3. If $\sigma = 0$ and $\int_{(-1,0)} |x| \varPi(\mathrm{d}x) < \infty$ then
$$W \in C^{k+1}(0, \infty) \Leftrightarrow \overline{\varPi} \in C^k(0, \infty).$$

On a final note we mention that Döring and Savov [37] have obtained further results concerning smoothness for potential measures of subordinators which have implications for the smoothness of scale functions. A particular result of interest in their paper is understanding where smoothness breaks down when atoms are introduced into the Lévy measure.

4 Engineering Scale Functions

4.1 Construction Through the Wiener–Hopf Factorization

Recall that the spatial Wiener–Hopf factorization can be expressed through the identity
$$\psi(\lambda) = (\lambda - \varPhi(0))\phi(\lambda), \tag{66}$$
for $\lambda \geq 0$ where $\phi(\lambda)$ is the Laplace exponent of the descending ladder height process. The killing rate, drift coefficient and Lévy measure associated with ϕ are given by $\psi'(0+) \vee 0$, $\sigma^2/2$ and
$$\varUpsilon(x, \infty) = \mathrm{e}^{\varPhi(0)x} \int_x^\infty \mathrm{e}^{-\varPhi(0)u} \varPi(-\infty, -u)\mathrm{d}u, \quad \text{for } x > 0,$$
respectively. Moreover, from this decomposition one may derive that
$$W(x) = \mathrm{e}^{\varPhi(0)x} \int_0^x \mathrm{e}^{-\varPhi(0)y} \int_0^\infty \mathrm{d}t \cdot P\left(\widehat{H}_t \in \mathrm{d}y\right), \quad x \geq 0. \tag{67}$$

This relationship between scale functions and potential measures of subordinators lies at the heart of the approach we shall describe in this section. Key to the method is the fact that one can find in the literature several subordinators for which the potential measure is known explicitly. Should these subordinators turn out to be the descending ladder height process of a spectrally negative Lévy process which does not drift to $-\infty$, i.e. $\Phi(0) = 0$, then this would give an exact expression for its scale function. Said another way, we can build scale functions using the following approach.

Step 1. Choose a subordinator, say $\widehat{H}$, with Laplace exponent ϕ, for which one knows its potential measure or equivalently, in light of (38), one can explicitly invert the Laplace transform $1/\phi(\theta)$.

Step 2. Verify whether the relation

$$\psi(\lambda) := \lambda\phi(\lambda), \qquad \lambda \geq 0,$$

defines the Laplace exponent of a spectrally negative Lévy process.

Of course, for this method to be useful we should first provide necessary and sufficient conditions for a subordinator to be the downward ladder height process of some spectrally negative Lévy process or equivalently a verification method for Step 2.

The following theorem, taken from Hubalek and Kyprianou [49] (see also Vigon [104]), shows how one may identify a spectrally negative Lévy process X (called the *parent process*) for a given descending ladder height process $\widehat{H}$. The proof follows by a straightforward manipulation of the Wiener–Hopf factorization (66).

Theorem 4.1. *There is a bijection between the class of spectrally negative Lévy processes and that of subordinators whose Lévy measure admits a non-increasing density. More precisely, suppose that $\widehat{H}$ is a subordinator, killed at rate $\kappa \geq 0$, with drift δ and Lévy measure Υ which is absolutely continuous with non-increasing density. Suppose further that $\varphi \geq 0$ is given such that $\varphi\kappa = 0$. Then there exists a spectrally negative Lévy process X, henceforth referred to as the "parent process," such that for all $x \geq 0$, $\mathbb{P}(\tau_x^+ < \infty) = e^{-\varphi x}$ and whose descending ladder height process is precisely the process $\widehat{H}$. The Lévy triple (a, σ, Π) of the parent process is uniquely identified as follows. The Gaussian coefficient is given by $\sigma = \sqrt{2\delta}$. The Lévy measure is given by*

$$\Pi(-\infty, -x) = \varphi\Upsilon(x, \infty) + \frac{d\Upsilon}{dx}(x). \tag{68}$$

Finally

$$a = \int_{(-\infty, -1)} x\Pi(dx) - \kappa \tag{69}$$

if $\varphi = 0$ and otherwise when $\varphi > 0$

$$a = \frac{1}{2}\sigma^2\varphi + \frac{1}{\varphi}\int_{(-\infty, 0)} (e^{\varphi x} - 1 - x\varphi\mathbf{1}_{\{x > -1\}})\Pi(dx). \tag{70}$$

In all cases, the Laplace exponent of the parent process is also given by

$$\psi(\theta) = (\theta - \varphi)\phi(\theta), \tag{71}$$

for $\theta \geq 0$ where $\phi(\theta) = -\log \mathbb{E}(e^{-\theta \widehat{H}_1})$.

Conversely, the killing rate, drift and Lévy measure of the descending ladder height process associated to a given spectrally negative Lévy process X satisfying $\varphi = \Phi(0)$ are also given by the above formulae.

Note that when describing parent processes later on in this text, for practical reasons, we shall prefer to specify the triple (σ, Π, ψ) instead of (a, σ, Π). However both triples provide an equivalent amount of information. It is also worth making an observation for later reference concerning the path variation of the process X for a given a descending ladder height process H.

Corollary 4.2. *Given a killed subordinator H satisfying the conditions of the previous theorem,*

(i) The parent process has paths of unbounded variation if and only if $\Upsilon(0, \infty) = \infty$ or $\delta > 0$,

(ii) If $\Upsilon(0, \infty) = \lambda < \infty$ then the parent process necessarily decomposes in the form

$$X_t = (\kappa + \lambda - \delta\varphi)t + \sqrt{2\delta}B_t - S_t, \tag{72}$$

where $B = \{B_t : t \geq 0\}$ is a Brownian motion, $S = \{S_t : t \geq 0\}$ is an independent driftless subordinator with Lévy measure ν satisfying

$$\nu(x, \infty) = \varphi\Upsilon(x, \infty) + \frac{\mathrm{d}\Upsilon}{\mathrm{d}x}(x)$$

for all $x > 0$.

Proof. The path variation of X follows directly from (68) and the fact that $\sigma = \sqrt{2\delta}$. Also using (68), the Laplace exponent of the decomposition (72) can be computed as follows with the help of an integration by parts;

$$(\kappa + \lambda - \delta\varphi)\theta + \delta\theta^2 - \varphi\theta \int_0^\infty e^{-\theta x}\Upsilon(x, \infty)\mathrm{d}x - \theta \int_0^\infty e^{-\theta x}\frac{\mathrm{d}\Upsilon}{\mathrm{d}x}(x)\mathrm{d}x$$

$$= (\kappa + \Upsilon(0, \infty) - \delta\varphi)\theta + \delta\theta^2 - \varphi \int_0^\infty (1 - e^{-\theta x})\frac{\mathrm{d}\Upsilon}{\mathrm{d}x}(x)\mathrm{d}x$$

$$- \theta \int_0^\infty e^{-\theta x}\frac{\mathrm{d}\Upsilon}{\mathrm{d}x}(x)\mathrm{d}x$$

$$= (\theta - \varphi)\left(\kappa + \delta\theta + \int_0^\infty (1 - e^{-\theta x})\frac{\mathrm{d}\Upsilon}{\mathrm{d}x}(x)\mathrm{d}x\right).$$

This agrees with the Laplace exponent $\psi(\theta) = (\theta - \varphi)\phi(\theta)$ of the parent process constructed in Theorem 4.1. □

Example 4.3. Consider a spectrally negative Lévy process which is the parent process of a (killed) tempered stable process. That is to say a subordinator with Laplace exponent given by

$$\phi(\theta) = \kappa - c\Gamma(-\alpha)\left((\gamma + \theta)^\alpha - \gamma^\alpha\right),$$

where $\alpha \in (-1, 1) \setminus \{0\}$, $\gamma \geq 0$ and $c > 0$. For $\alpha, \beta > 0$, we will denote by

$$\mathscr{E}_{\alpha,\beta}(x) = \sum_{n \geq 0} \frac{x^n}{\Gamma(n\alpha + \beta)}, \qquad x \in \mathbb{R},$$

the two parameter Mittag-Leffler function. It is characterised by a pseudo-Laplace transform. Namely, for $\lambda \in \mathbb{R}$ and $\Re(\theta) > \lambda^{1/\alpha} - \gamma$,

$$\int_0^\infty e^{-\theta x} e^{-\gamma x} x^{\beta-1} \mathscr{E}_{\alpha,\beta}(\lambda x^\alpha) dx = \frac{(\theta + \gamma)^{\alpha-\beta}}{(\theta + \gamma)^\alpha - \lambda}. \tag{73}$$

One easily deduces the following transformations as special examples of (73) for $\theta, \lambda > 0$,

$$\int_0^\infty e^{-\theta x} x^{\alpha-1} \mathscr{E}_{\alpha,\alpha}(\lambda x^\alpha) dx = \frac{1}{\theta^\alpha - \lambda}, \tag{74}$$

and

$$\int_0^\infty e^{-\theta x} \lambda^{-1} x^{-\alpha-1} \mathscr{E}_{-\alpha,-\alpha}(\lambda^{-1} x^{-\alpha}) dx = \frac{\lambda}{\lambda - \theta^\alpha} - 1, \tag{75}$$

valid for $\alpha > 0$, resp. $\alpha < 0$. Together with the well-known rules for Laplace transforms concerning primitives and tilting allow us to quickly deduce the following expressions for the scale functions associated to the parent process with Laplace exponent given by (71) such that $\kappa\varphi = 0$.

If $0 < \alpha < 1$ then

$$W(x) = \frac{e^{\varphi x}}{-c\Gamma(-\alpha)} \int_0^x e^{-(\gamma+\varphi)y} y^{\alpha-1} \mathscr{E}_{\alpha,\alpha}\left(\frac{\kappa + c\Gamma(-\alpha)\gamma^\alpha}{c\Gamma(-\alpha)} y^\alpha\right) dy.$$

If $-1 < \alpha < 0$, then

$$W(x) = \frac{e^{\varphi x}}{\kappa + c\Gamma(-\alpha)\gamma^\alpha}$$

$$+ \frac{c\Gamma(-\alpha)e^{\varphi x}}{(\kappa + c\Gamma(-\alpha)\gamma^\alpha)^2} \int_0^x e^{-(\gamma+\varphi)y} y^{-\alpha-1} \mathscr{E}_{-\alpha,-\alpha}\left(\frac{c\Gamma(-\alpha)y^{-\alpha}}{\kappa + c\Gamma(-\alpha)\gamma^\alpha}\right) dy.$$

Example 4.4. Let $c > 0$, $\nu \geq 0$ and $\theta \in (0, 1)$ and ϕ be defined by

$$\phi(\lambda) = \frac{c\lambda\Gamma(\nu + \lambda)}{\Gamma(\nu + \lambda + \theta)}, \qquad \lambda \geq 0.$$

Elementary but tedious calculations using the Beta integral allow to prove that ϕ is the Laplace exponent of some subordinator, $\widehat{H}$. Its characteristics are $\kappa = 0$, $\delta = 0$,

$$\overline{\Upsilon}(x) := \Upsilon(x, \infty) = \frac{c}{\Gamma(\theta)} e^{-x(\nu + \theta - 1)} \left(e^x - 1\right)^{\theta - 1}, \qquad x > 0.$$

Moreover, we have that $\overline{\Pi}_{\widehat{H}}$ is non-increasing and log-convex, so $\Pi_{\widehat{H}}$ has a non-increasing density. It follows from Theorem 4.1 that there exists an oscillating spectrally negative Lévy process, say X, whose Laplace exponent is $\psi(\lambda) = \lambda\phi(\lambda)$, $\lambda \geq 0$, with $\sigma = 0$, and Lévy density given by $-\mathrm{d}^2\overline{\Upsilon}/\mathrm{d}x^2$. Using again the Beta integral we can obtain the potential measure of $\widehat{H}$, and as a consequence the scale function associated to X, which turns out to be given by

$$W(x) = \frac{\Gamma(\nu + \theta)}{c\Gamma(\nu)} + \frac{\theta}{c\Gamma(1 - \theta)} \int_0^x \left\{ \int_y^\infty \frac{e^{z(1-\nu)}}{(e^z - 1)^{1+\theta}} \mathrm{d}z \right\} \mathrm{d}y, \qquad x \geq 0.$$

The integral that defines this scale function can be calculated using the hypergeometric series. The particular case where $\nu = 1$, $c = \Gamma(1+\theta)$ appears in Chaumont et al. [31] and Patie [80].

An interesting feature of this example is that it comes together with another example. Indeed, observe that the first derivative of W is given by,

$$W'(x) = \int_x^\infty \frac{e^{z(1-\nu)}}{(e^z - 1)^{1+\theta}} \mathrm{d}z, \qquad x \geq 0,$$

which is non-increasing, convex and such that the second derivative satisfies the integrability condition $\int_0^\infty (1 \wedge x)|W''(x)|\mathrm{d}x < \infty$. So, $|W''(x)|$ defines the Lévy density of some subordinator. More precisely, the function defined by

$$\phi^*(\lambda) := \frac{\lambda}{\phi(\lambda)}$$

$$= \lambda \int_0^\infty e^{-\lambda x} dW(x)$$

$$= \frac{\Gamma(\nu + \theta)}{c\Gamma(\nu)} + \frac{\theta}{c\Gamma(1 - \theta)} \int_0^\infty (1 - e^{-\lambda x}) \frac{e^{x(1-\nu)}}{(e^x - 1)^{1+\theta}} \mathrm{d}x, \ \lambda \geq 0,$$

is the Laplace exponent of some subordinator H^*, which in turn has a Lévy measure with a non-increasing density. Hence

$$\psi^*(\lambda) = \lambda\phi^*(\lambda) = \frac{\lambda^2}{\phi(\lambda)}, \qquad \lambda \geq 0,$$

defines the Laplace exponent of a spectrally negative Lévy process that drifts to ∞. It can be easily verified by an integration by parts that the associated scale function is given by

$$W^*(x) = \frac{c}{\Gamma(\theta)} \int_0^x e^{-z(\nu+\theta-1)}(e^z - 1)^{\theta-1}dz = \int_0^x \overline{\Upsilon}(z)dz, \quad x \geq 0.$$

The method described in the previous example for generating two examples of scale functions simultaneously can be formalized into a general theory that applies to a large family of subordinators, namely that of special subordinators.

4.2 Special and Conjugate Scale Functions

In this section we introduce the notion of a special Bernstein functions and special subordinators and use the latter to justify the existence of pairs of so called *conjugate scale functions* which have a particular analytical structure. We refer the reader to the lecture notes of Song and Vondraček [95], the recent book by Schilling et al. [92] and the books of Berg and Gunnar [13] and Jacobs [51] for a more complete account of the theory of Bernstein functions and their application in potential analysis.

Recall that the class of Bernstein functions coincides precisely with the class of Laplace exponents of possibly killed subordinators. That is to say, a general Bernstein function takes the form

$$\phi(\theta) = \kappa + \delta\theta + \int_{(0,\infty)} (1 - e^{-\theta x})\Upsilon(dx), \text{ for } \theta \geq 0, \tag{76}$$

where $\kappa \geq 0$, $\delta \geq 0$ and Υ is a measure concentrated on $(0, \infty)$ such that $\int_{(0,\infty)} (1 \wedge x)\Upsilon(dx) < \infty$.

Definition 4.5. Suppose that $\phi(\theta)$ is a Bernstein function, then it is called a *special Bernstein function* if

$$\phi(\theta) = \frac{\theta}{\phi^*(\theta)}, \qquad \theta \geq 0, \tag{77}$$

where $\phi^*(\theta)$ is another Bernstein function. In this case we will say that the Bernstein function ϕ^* is conjugate to ϕ. Accordingly a possibly killed subordinator is called a special subordinator if its Laplace exponent is a special Bernstein function.

It is apparent from its definition that ϕ^* is a special Bernstein function and ϕ is its conjugate. In [47] and [96] it is shown that a sufficient condition for ϕ to be a special subordinator is that $\Upsilon(x, \infty)$ is log-convex on $(0, \infty)$.

For conjugate pairs of special Bernstein functions ϕ and ϕ^* we shall write in addition to (77)

$$\phi^*(\theta) = \kappa^* + \delta^*\theta + \int_{(0,\infty)} (1 - e^{-\theta x})\Upsilon^*(dx), \qquad \theta \geq 0, \qquad (78)$$

where necessarily Υ^* is a measure concentrated on $(0,\infty)$ satisfying $\int_{(0,\infty)}(1 \wedge x)$ $\Upsilon^*(dx) < \infty$. One may express the triple $(\kappa^*, \delta^*, \Upsilon^*)$ in terms of related quantities coming from the conjugate ϕ. Indeed it is known that

$$\kappa^* = \begin{cases} 0, & \kappa > 0, \\ \left(\delta + \int_{(0,\infty)} x\Upsilon(dx)\right)^{-1}, & \kappa = 0; \end{cases}$$

and

$$\delta^* = \begin{cases} 0, & \delta > 0 \text{ or } \Upsilon(0,\infty) = \infty, \\ (\kappa + \Upsilon(0,\infty))^{-1}, & \delta = 0 \text{ and } \Upsilon(0,\infty) < \infty. \end{cases} \qquad (79)$$

Which implies in particular that $\kappa^*\kappa = 0 = \delta^*\delta$. In order to describe the measure Υ^* let us denote by $W(dx)$ the potential measure of ϕ. (This choice of notation is of course pre-emptive.) Then we have that W necessarily satisfies

$$W(dx) = \delta^*\delta_0(dx) + \{\kappa^* + \Upsilon^*(x,\infty)\}\,dx, \qquad \text{for } x \geq 0,$$

where $\delta_0(dx)$ is the Dirac measure at zero. Naturally, if W^* is the potential measure of ϕ^* then we may equally describe it in terms of $(\kappa, \delta, \Upsilon)$. In fact it can be easily shown that a necessary and sufficient condition for a Bernstein function to be special is that its potential measure has a density on $(0,\infty)$ which is non-increasing and integrable in the neighborhood of the origin.

We are interested in constructing a parent process whose descending ladder height process is a special subordinator. The following theorem and corollary are now evident given the discussion in the current and previous sections.

Theorem 4.6. *For conjugate special Bernstein functions ϕ and ϕ^* satisfying (76) and (78) respectively, where Υ is absolutely continuous with non-increasing density, there exists a spectrally negative Lévy process that does not drift to $-\infty$, whose Laplace exponent is described by*

$$\psi(\theta) = \frac{\theta^2}{\phi^*(\theta)} = \theta\phi(\theta), \text{ for } \theta \geq 0, \qquad (80)$$

and whose scale function is a concave function and is given by

$$W(x) = \delta^* + \kappa^*x + \int_0^x \Upsilon^*(y,\infty)dy. \qquad (81)$$

The assumptions of the previous theorem require only that the Lévy and potential measures associated to ϕ have a non-increasing density in $(0, \infty)$, respectively; this condition on the potential measure is equivalent to the existence of ϕ^*. If in addition it is assumed that the potential density be a convex function, in light of the representation (81), we can interchange the roles of ϕ and ϕ^*, respectively, in the previous theorem. The key issue to this additional assumption is that it ensures the absolute continuity of Υ^* with a non-increasing density and hence that we can apply the Theorem 4.1. We thus have the following Corollary.

Corollary 4.7. *If conjugate special Bernstein functions ϕ and ϕ^* exist satisfying (76) and (78) such that both Υ and Υ^* are absolutely continuous with non-increasing densities, then there exist a pair of scale functions W and W^*, such that W is concave, its first derivative is a convex function, (81) is satisfied, and*

$$W^*(x) = \delta + \kappa x + \int_0^x \Upsilon(y, \infty) dy. \tag{82}$$

Moreover, the respective parent processes are given by (80) and

$$\psi^*(\theta) = \frac{\theta^2}{\phi(\theta)} = \theta \phi^*(\theta). \tag{83}$$

It is important to mention that the converse of the previous Theorem and Corollary hold true but we omit a statement and proof for sake of brevity and refer the reader to the article [68] for further details.

For obvious reasons we shall henceforth refer to the scale functions identified in (76) and (78) as *special scale functions*. Similarly, when W and W^* exist then we refer to them as *conjugate (special) scale functions* and their respective parent processes are called *conjugate parent processes*. This conjugation can be seen by noting that thanks to (77), on $x \geq 0$,

$$W * W^*(dx) = dx.$$

4.3   Tilting and Parent Processes Drifting to $-\infty$

In this section we present two methods for which, given a scale function and associated parent process, it is possible to construct further examples of scale functions by appealing to two procedures.

The first method relies on the following known facts concerning translating the argument of a given Bernstein function. Let ϕ be a special Bernstein function with representation given by (76). Then for any $\beta \geq 0$ the function $\phi_\beta(\theta) = \phi(\theta + \beta)$, $\theta \geq 0$, is also a special Bernstein function with killing term $\kappa_\beta = \phi(\beta)$, drift term $d_\beta = d$ and Lévy measure $\Upsilon_\beta(dx) = e^{-\beta x} \Upsilon(dx)$, $x > 0$, see e.g. [91] Sect. 33.

Its associated potential measure, W_β, has a decreasing density in $(0, \infty)$ such that $W_\beta(\mathrm{d}x) = \mathrm{e}^{-\beta x}W'(x)\mathrm{d}x$, $x > 0$, where W' denotes the density of the potential measure associated to ϕ. Moreover, let ϕ^* and ϕ^*_β, denote the conjugate Bernstein functions of ϕ and ϕ_β, respectively. Then the following identity

$$\phi^*_\beta(\theta) = \phi^*(\theta+\beta) - \phi^*(\beta) + \beta \int_0^\infty \left(1 - \mathrm{e}^{-\theta x}\right)\mathrm{e}^{-\beta x}W'(x)\mathrm{d}x, \qquad \theta \geq 0, \quad (84)$$

holds. Note in particular that if Υ has a non-increasing density then so does Υ_β. Moreover, if W' is convex (equivalently Υ^* has a non-increasing density) then W'_β is convex (equivalently Υ^*_β has a non-increasing density). These facts lead us to the following Lemma.

Lemma 4.8. *If conjugate special Bernstein functions ϕ and ϕ^* exist satisfying (76) and (78) such that both Υ and Υ^* are absolutely continuous with non-increasing densities, then there exist conjugate parent processes with Laplace exponents*

$$\psi_\beta(\theta) = \theta\phi_\beta(\theta) \text{ and } \psi^*_\beta(\theta) = \theta\phi^*_\beta(\theta), \qquad \theta \geq 0,$$

whose respective scale functions are given by

$$W_\beta(x) = \delta^* + \int_0^x \mathrm{e}^{-\beta y}(\kappa + \Upsilon^*(y, \infty))\mathrm{d}y$$
$$= \mathrm{e}^{-\beta x}W(x) + \beta \int_0^x \mathrm{e}^{-\beta z}W(z)\mathrm{d}z, \qquad x \geq 0, \quad (85)$$

and

$$W^*_\beta(x) = \delta + \phi(\beta)x + \int_0^x \left(\int_y^\infty \mathrm{e}^{-\beta z}\Upsilon(\mathrm{d}z)\right)\mathrm{d}y, \qquad (86)$$

where we have used obvious notation.

We would like to stress that the statement in the first line of (85) corrects a typographical error in Lemma 1 in [68]. All the statements in this Lemma, except the equality (85), follow from the previous discussion. The equality (85) is a simple consequence of the expression of W in (81) and an integration by parts.

The second method builds on the first to construct examples of scale functions whose parent processes may be seen as an auxiliary parent process conditioned to drift to $-\infty$.

Suppose that ϕ is a Bernstein function such that $\phi(0) = 0$, its associated Lévy measure has a decreasing density and let $\beta > 0$. Theorem 4.1, as stated in its more general form in [49], says that there exists a parent process, say X, that drifts to $-\infty$ such that its Laplace exponent ψ can be factorized as

$$\psi(\theta) = (\theta - \beta)\phi(\theta), \qquad \theta \geq 0.$$

It follows that ψ is a convex function and $\psi(0) = 0 = \psi(\beta)$, so that β is the largest positive solution to the equation $\psi(\theta) = 0$. Now, let W_β be the 0-scale function of the spectrally negative Lévy process, say X_β, with Laplace exponent $\psi_\beta(\theta) := \psi(\theta + \beta)$, for $\theta \geq 0$. It is known that the Lévy process X_β is obtained by an exponential change of measure and can be seen as the Lévy process X conditioned to drift to ∞, see Chap. VII in [15]. Thus the Laplace exponent ψ_β can be factorized as $\psi_\beta(\theta) = \theta\phi_\beta(\theta)$, for $\theta \geq 0$, where, as before, $\phi_\beta(\cdot) := \phi(\beta + \cdot)$. It follows from Lemma 8.4 in [66], that the 0-scale function of the process with Laplace exponent ψ is related to W_β by

$$W(x) = e^{\beta x} W_\beta(x), \qquad x \geq 0.$$

The above considerations thus lead to the following result which allows for the construction of a second parent process and associated scale function over and above the pair described in Theorem 4.6.

Lemma 4.9. *Suppose that ϕ is a special Bernstein function satisfying (76) such that Υ is absolutely continuous with non-increasing density and $\kappa = 0$. Fix $\beta > 0$. Then there exists a parent process with Laplace exponent*

$$\psi(\theta) = (\theta - \beta)\phi(\theta), \quad \theta \geq 0,$$

whose associated scale function is given by

$$W(x) = \delta^* e^{\beta x} + e^{\beta x} \int_0^x e^{-\beta y} \Upsilon^*(y, \infty) dy, \qquad x \geq 0,$$

where we have used our usual notation.

In [68] the interested reader may find a discussion about the conjugated pairs arising in this construction. Note that the conclusion of this Lemma can also be derived from (40) and (81).

4.4   Complete Scale Functions

We have seen several methods that allow us to construct scale functions and pairs of conjugate scale functions which in principle generate large families of scale functions. In particular the method of constructing pairs of conjugate scale functions and their tilted versions needs the hypothesis of decreasing densities for the Lévy measures of the underlying conjugate subordinators. This may be a serious issue because in order to verify that hypothesis one needs to determine explicitly both densities, which can be a very hard and technical task. Luckily, there is a large class of Bernstein functions, the so-called *complete Bernstein functions*, for which this condition is satisfied automatically. Hence, our purpose in this section is to recall some of the keys facts related to this class and its consequences.

We begin by introducing the notion of a complete Bernstein function with a view to constructing scale functions whose parent processes are derived from descending ladder height processes with Laplace exponents which belong to the class of complete Bernstein functions.

Definition 4.10. A function ϕ is called *complete Bernstein function* if there exists an auxiliary Bernstein function η such that

$$\phi(\theta) = \theta^2 \int_{(0,\infty)} e^{-\theta x} \eta(x) \mathrm{d}x. \tag{87}$$

It is well known that a complete Bernstein function is necessarily a special Bernstein function (cf. [51]) and in addition, its conjugate is also a complete Bernstein function. Moreover, from the same reference one finds that a necessary and sufficient condition for ϕ to be complete Bernstein is that Υ satisfies for $x > 0$

$$\Upsilon(\mathrm{d}x) = \left\{ \int_{(0,\infty)} e^{-xy} \gamma(\mathrm{d}y) \right\} \mathrm{d}x,$$

where $\int_{(0,1)} \frac{1}{y} \gamma(\mathrm{d}y) + \int_{(1,\infty)} \frac{1}{y^2} \gamma(\mathrm{d}y) < \infty$. Equivalently Υ has a completely monotone density. Another necessary and sufficient condition is that the potential measure associated to ϕ has a density on $(0, \infty)$ which is completely monotone, this is a result due to Kingman [54] and Hawkes [48]. The class of infinitely divisible laws and subordinators related to this type of Bernstein functions has been extensively studied by several authors, see e.g. [23,35,52,90,102] and the references therein.

Since Υ is necessarily absolutely continuous with a completely monotone density, it follows that any subordinator whose Laplace exponent is a complete Bernstein function may be used in conjunction with Corollary 4.7. The following result is now a straightforward application of the latter and the fact that from (87), any Bernstein function η has a Laplace transform $\phi(\theta)/\theta^2$ where ϕ is complete Bernstein.

Corollary 4.11. *Let η be any Bernstein function and suppose that ϕ is the complete Bernstein function associated with the latter via the relation (87). Write ϕ^* for the conjugate of ϕ and η^* for the Bernstein function associated with ϕ^* via the relation (87). Then*

$$W(x) = \eta^*(x) \text{ and } W^*(x) = \eta(x), \qquad x \geq 0,$$

are conjugate scale functions with conjugate parent processes whose Laplace exponents are given by

$$\psi(\theta) = \frac{\theta^2}{\phi^*(\theta)} = \theta \phi(\theta) \text{ and } \psi^*(\theta) = \frac{\theta^2}{\phi(\theta)} = \theta \phi^*(\theta), \qquad \theta \geq 0.$$

Again, for obvious reasons, we shall refer to the scale functions described in the above corollary as *complete scale functions*. Note that W is a complete scale function if and only if the descending ladder height subordinator has a completely monotone density.

An interesting and useful consequence of this result is that any given Bernstein function η is a scale function whose parent process is the spectrally negative Lévy process whose Laplace exponent is given by $\psi^*(\theta) = \theta^2/\phi(\theta)$ where ϕ is given by (87). Another interesting and straightforward consequence we capture in the corollary below. Examples follow.

Corollary 4.12. *If W is any complete scale function then so is*

$$a + bx + W(x), \qquad x \geq 0,$$

for any $a, b \geq 0$.

Example 4.13. Let $0 < \alpha < \beta \leq 1$, $a, b > 0$ and ϕ be the Bernstein function defined by

$$\phi(\theta) = a\theta^{\beta-\alpha} + b\theta^\beta, \qquad \theta \geq 0.$$

That is, in the case where $\alpha < \beta < 1$, ϕ is the Laplace exponent of a subordinator which is obtained as the sum of two independent stable subordinators one of parameter $\beta - \alpha$ and the other of parameter β, respectively, so that the killing and drift term of ϕ are both equal to 0, and its Lévy measure is given by

$$\Upsilon(\mathrm{d}x) = \left(\frac{a(\beta - \alpha)}{\Gamma(1 - \beta + \alpha)} x^{-(1+\beta-\alpha)} + \frac{b\beta}{\Gamma(1 - \beta)} x^{-(1+\beta)} \right) \mathrm{d}x, \qquad x > 0.$$

In the case that $\beta = 1$, ϕ is the Laplace exponent of a stable subordinator with parameter $1 - \alpha$ and a linear drift. In all cases, the underlying Lévy measure has a density which is completely monotone, and thus its potential density, or equivalently the density of the associated scale function W, is completely monotone.

We recall that the two parameter Mittag-Leffler function is defined by

$$\mathscr{E}_{\alpha,\beta}(x) = \sum_{n \geq 0} \frac{x^n}{\Gamma(n\alpha + \beta)}, \qquad x \in \mathbb{R}, \tag{88}$$

where $\alpha, \beta > 0$. With the help of the transformation (73), the associated scale function to ϕ can now be identified via

$$W'(x) = \frac{1}{b} x^{\beta-1} \mathscr{E}_{\alpha,\beta}\left(-ax^\alpha/b\right), \qquad x > 0, \tag{89}$$

which is a completely monotone function. Hence

$$\psi(\theta) = \theta\phi(\theta) = a\theta^{\beta-\alpha+1} + b\theta^{\beta+1}, \qquad \theta \geq 0,$$

is the Laplace exponent of a spectrally negative Lévy process, the parent process. It oscillates and is obtained by adding two independent spectrally negative stable processes with stability index $\beta + 1$ and $1 + \beta - \alpha$, respectively. The scale function associated to it is given by

$$W(x) = \frac{1}{b} \int_0^x t^{\beta-1} \mathscr{E}_{\alpha,\beta}(-at^\alpha/b) \mathrm{d}t, \qquad x \geq 0.$$

The associated conjugates are given by

$$\phi^*(\theta) = \frac{\theta}{a\theta^{\beta-\alpha} + b\theta^\beta}, \quad \psi^*(\theta) = \frac{\theta^2}{a\theta^{\beta-\alpha} + b\theta^\beta}, \qquad \theta \geq 0,$$

and

$$W^*(x) = \frac{a}{\Gamma(2-\beta+\alpha)} x^{1-\beta+\alpha} + \frac{b}{\Gamma(2-\beta)} x^{1-\beta}, \qquad x \geq 0. \qquad (90)$$

The subordinator with Laplace exponent ϕ^* has zero killing and drift terms and its Lévy measure is obtained by taking the derivative of the expression in (89). By Theorem 4.1 the spectrally negative Lévy process with Laplace exponent ψ^*, oscillates, has unbounded variation, has zero Gaussian term, and its Lévy measure is obtained by taking the second derivative of the expression in (89).

One may mention here that by letting $a \downarrow 0$ the Continuity Theorem for Laplace transforms tells us that for the case $\phi(\theta) = b\theta^\beta$, the associated ψ is the Laplace exponent of a spectrally negative stable process with stability parameter $1 + \beta$, and its scale function is given by

$$W(x) = \frac{1}{b\Gamma(1+\beta)} x^\beta, \qquad x \geq 0.$$

The associated conjugates are given by

$$\phi^*(\theta) = b^{-1}\theta^{1-\beta}, \quad \psi^*(\theta) = b^{-1}\theta^{2-\beta}, \qquad \theta \geq 0,$$

and

$$W^*(x) = \frac{b}{\Gamma(2-\beta)} x^{1-\beta}, \quad x \geq 0.$$

So that ϕ^* (respectively ψ^*) corresponds to a stable subordinator of parameter $1-\beta$, zero killing and drift terms (respectively, to a oscillating spectrally negative stable Lévy process with stability index $2 - \beta$), and so its Lévy measure is given by

$$\Pi^*(-\infty, -x) = \frac{\beta(1-\beta)}{b\Gamma(1+\beta)} x^{\beta-2}, \qquad x \geq 0.$$

To complete this example, observe that the change of measure introduced in Lemma 4.8 allows us to deal with the Bernstein function

$$\phi(\theta) = k(\theta + m)^{\beta - \alpha} + b(\theta + m)^{\beta}, \quad \theta \geq 0,$$

where $m \geq 0$ is a fixed parameter. In this case we get that there exists a spectrally negative Lévy process whose Laplace exponent is given by

$$\psi(\theta) = k\theta(\theta + m)^{\beta - \alpha} + b\theta(\theta + m)^{\beta}, \quad \theta \geq 0,$$

and its associated scale function is given by

$$W(x) = \frac{1}{b} \int_0^x e^{-mt} t^{\beta - 1} \mathscr{E}_{\alpha, \beta}(-at^{\alpha}/b) dt, \qquad x \geq 0.$$

The respective conjugates can be obtained explicitly but we omit the details given that the expressions found are too involved.

We can now use the construction in Sect. 4.3. For $m, a, b > 0, 0 < \alpha < \beta < 1$, there exists a parent process drifting to $-\infty$ and with Laplace exponent

$$\psi(\theta) = (\theta - m)\left(a\theta^{\beta - \alpha} + b\theta^{\beta}\right), \qquad \theta \geq 0.$$

It follows from the previous calculations that the scale function associated to the parent process with Laplace exponent ψ is given by

$$W(x) = \frac{e^{mx}}{b} \int_0^x e^{-mt} t^{\beta - 1} \mathscr{E}_{\alpha, \beta}(-at^{\alpha}/b) dt, \qquad x \geq 0.$$

Example 4.14. Consider the case of a spectrally negative stable process with index $\gamma \in (1, 2)$. Its Laplace exponent is given by $\psi(\theta) = \theta^{\gamma}$ and its scale function is given by $W(x) = x^{\gamma - 1}/\Gamma(\gamma)$. Corollary 4.12 predicts that $1 + x^{\gamma - 1}/\Gamma(\gamma), x \geq 0$, is a scale function of a process whose Laplace exponent is given by

$$\left(\frac{1}{\theta} + \frac{1}{\theta^{\gamma}}\right)^{-1} = \frac{\theta^2}{\theta^{2 - \gamma} + \theta}, \qquad \theta \geq 0.$$

Note that this example can also be recovered from (90) with an appropriate choice of constants.

4.5 Generating Scale Functions via an Analytical Transformation

In Chazal et al. [32] it was shown that whenever ψ is the Laplace exponent of a spectrally negative Lévy process then so is

$$\mathscr{T}_{\delta, \beta} \psi^{(q)}(u) := \frac{u + \beta - \delta}{u + \beta} \psi^{(q)}(u + \beta) - \frac{\beta - \delta}{\beta} \psi^{(q)}(\beta), \quad u \geq -\beta, \, u \geq 0,$$

where $\psi^{(q)}(u) = \psi(u) - q$, $\delta, \beta \geq 0$ and $\psi^{(q)\prime}(0+) = q = 0$ if $\beta = 0 < \delta$. When $\beta = \delta = 0$ we understand $\mathscr{T}_{0,0}\psi^{(q)}(u) = \psi^{(q)}(u)$. Note that $\psi^{(q)}$ is the Laplace exponent of a spectrally negative Lévy process possibly killed at an independent and exponentially distributed random time with rate $q \geq 0$. In the usual way, when $q = 0$ we understand there to be no killing. Moreover, the characteristics of the Lévy process with Laplace exponent $\mathscr{T}_{\delta,\beta}\psi$ are also described in the aforementioned paper through the following result.

Proposition 4.15. *Fix $\delta, \beta \geq 0$ with the additional constraint that $\psi^{(q)\prime}(0+) = q = 0$ if $\beta = 0 < \delta$. If $\psi^{(q)}$ has Gaussian coefficient σ and jump measure Π then $\mathscr{T}_{\delta,\beta}\psi^{(q)}$ also has Gaussian coefficient σ and its Lévy measure is given by*

$$e^{\beta x}\Pi(\mathrm{d}x) + \delta e^{\beta x}\overline{\Pi}(x)\mathrm{d}x + \delta\frac{\kappa}{\beta}e^{\beta x}\mathrm{d}x \quad on \ (-\infty, 0),$$

where $\overline{\Pi}(x) = \Pi(-\infty, -x)$.

In particular we note that $\mathscr{T}_{\delta,\beta}\psi^{(q)}$ is the Laplace exponent of a spectrally negative Lévy process without killing, i.e. $\mathscr{T}_{\delta,\beta}\psi^{(q)}(0) = 0$.

It turns out that this transformation can be used in a very straightforward way in combination with the definition of q-scale functions to generate new examples. Note for example that for a given ψ and $q \geq 0$ we have

$$\int_0^\infty e^{-\theta x}W^{(q)}(x)\mathrm{d}x = \frac{1}{\psi^{(q)}(\theta)}, \quad \text{for } \theta > \Phi(q),$$

and hence it is natural to use this transformation to help find the Laplace inverse (if it exists) of

$$\frac{1}{\mathscr{T}_{\delta,\beta}\psi^{(q)}(\theta)},$$

for β sufficiently large to give an expression for $W_{\mathscr{T}_{\beta,\delta}\psi^{(q)}}$, the 0-scale function associated with the spectrally negative Lévy process whose exponent is $\mathscr{T}_{\beta,\delta}\psi^{(q)}$. The following result, taken from [32] does exactly this. Note that the first conclusion in the theorem gives a similar result to the conclusion of Lemma 4.8 without the need for the descending ladder height to be special.

Theorem 4.16. *Let $x, \beta \geq 0$ such that $\psi'(0+) = q = 0$ if $\beta = 0 < \delta$. Then,*

$$W_{\mathscr{T}_{\beta,\beta}\psi^{(q)}}(x) = e^{-\beta x}W^{(q)}(x) + \beta\int_0^x e^{-\beta y}W^{(q)}(y)\mathrm{d}y. \tag{91}$$

Moreover, if $\psi'(0+) \leq 0$, then for any $x, \delta, q \geq 0$ we have

$$W_{\mathscr{T}_{\delta,\Phi(0)}\psi^{(q)}}(x) = e^{-\Phi(0)x}\left(W^{(q)}(x) + \delta e^{\delta x}\int_0^x e^{-\delta y}W^{(q)}(y)\mathrm{d}y\right).$$

Proof. The first assertion is proved by observing that

$$\int_0^\infty e^{-\theta x} W_{\mathscr{T}_{\beta,\beta}\psi^{(q)}}(x)\mathrm{d}x = \frac{\theta+\beta}{\theta\psi(\theta+\beta)}$$

$$= \frac{1}{\psi(\theta+\beta)} + \frac{\beta}{\theta\psi(\theta+\beta)},$$

which agrees with the Laplace transform of the right hand side of (91).

For the second claim, first note that $\mathscr{T}_{\delta,\theta}\psi^{(q)} = (\theta + \Phi(0) - \delta)\psi(\theta + \Phi(0))/(\theta + \Phi(0))$. A straightforward calculation shows that for all $\theta + \delta > \Phi(0)$, we have

$$\int_0^\infty e^{-\theta x} e^{(\Phi(0)-\delta)x} W_{\mathscr{T}_{\delta,\Phi(0)}\psi^{(q)}}(x)\mathrm{d}x = \frac{\theta+\delta}{\theta\psi(\theta+\delta)}.$$

The result now follows from the first part of the theorem.   □

Below we give an example of how this theory can be easily applied to generate new scale functions from those of (tempered) scale functions.

Example 4.17. Let

$$\psi_c^{(q)}(\theta) = (\theta+c)^\alpha - c^\alpha - q \text{ for } \theta \geq -c,$$

where $1 < \alpha < 2$ and $q, c \geq 0$. This is the Laplace exponent of an unbounded variation tempered stable spectrally negative Lévy process ξ killed at an independent and exponentially distributed time with rate q. In the case that $c = 0$, the underlying Lévy process is a spectrally negative α-stable Lévy process. In that case it is known that

$$\int_0^\infty e^{-\theta x} x^{\alpha-1} \mathscr{E}_{\alpha,\alpha}(qx^\alpha)\mathrm{d}x = \frac{1}{\theta^\alpha - q},$$

and hence the scale function is given by

$$W_{\psi_0^{(q)}}(x) = x^{\alpha-1}\mathscr{E}_{\alpha,\alpha}(qx^\alpha),$$

for $x \geq 0$. (Note in particular that when $q = 0$ the expression for the scale function simplifies to $\Gamma(\alpha)^{-1}x^{\alpha-1}$). Since

$$\int_0^\infty e^{-\theta x} e^{-cx} W_{\psi_0^{(q+c^\alpha)}}(x)\mathrm{d}x = \frac{1}{(\theta+c)^\alpha - c^\alpha - q},$$

it follows that

$$W_{\psi_c^{(q)}}(x) = e^{-cx} W_{\psi_0^{(q+c^\alpha)}}(x) = e^{-cx} x^{\alpha-1}\mathscr{E}_{\alpha,\alpha}((q+c^\alpha)x^\alpha).$$

Appealing to the first part of Theorem 4.16 we now know that for $\beta \geq 0$,

$$W_{\mathcal{T}_\beta \psi_c^{(q)}}(x) = e^{-(\beta+c)x}x^{\alpha-1}\mathscr{E}_{\alpha,\alpha}((q+c^\alpha)x^\alpha)$$

$$+ \beta \int_0^x e^{-(\beta+c)y}y^{\alpha-1}\mathscr{E}_{\alpha,\alpha}((q+c^\alpha)y^\alpha)dy.$$

Note that $\psi_c^{(q)\prime}(0+) = \alpha c^{\alpha-1}$ which is zero if and only if $c = 0$. We may use the second and third part of Theorem 4.16 in this case. Hence, for any $\delta > 0$, the scale function of the spectrally negative Lévy process with Laplace exponent $\mathcal{T}_{\delta,0}\psi_0^{(0)}$ is

$$W_{\mathcal{T}_{\delta,0}\psi_0^{(0)}}(x) = \frac{1}{\Gamma(\alpha-1)}e^{\delta x}\int_0^x e^{-\delta y}y^{\alpha-2}dy$$

$$= \frac{\delta^{\alpha-1}}{\Gamma(\alpha-1)}e^{\delta x}\Gamma(\alpha-1,\delta x),$$

where we have used the recurrence relation for the Gamma function and $\Gamma(a,b)$ stands for the incomplete Gamma function of parameters $a, b > 0$. Moreover, we have, for any $\beta > 0$,

$$W_{\mathcal{T}_{\delta,0}^\beta\psi_0^{(0)}}(x) = \frac{1}{\Gamma(\alpha-1)}\left(\frac{\beta^\alpha}{\beta-\delta}\Gamma(\alpha-1,\beta x) - e^{(\beta-\delta)x}\frac{\delta^\alpha}{\beta-\delta}\Gamma(\alpha-1,\delta x)\right).$$

Finally, the scale function of the spectrally negative Lévy process with Laplace exponent $\mathcal{T}_{\delta,0}^\beta\psi_0^{(q)}$ is given by

$$W_{\mathcal{T}_{\delta,0}^\beta\psi_0^{(q)}}(x) = \frac{\beta}{\beta-\delta}(x/\beta)^{\alpha-1}\mathscr{E}_{\alpha,\alpha-1}\left(x;\frac{q}{\beta}\right)$$

$$- \frac{\delta}{\beta-\delta}e^{-(\beta-\delta)x}(x/\delta)^{\alpha-1}\mathscr{E}_{\alpha,\alpha-1}\left(x;\frac{q}{\delta}\right),$$

where we have used the notation

$$\mathscr{E}_{\alpha,\beta}(x;q) = \sum_{n=0}^\infty \frac{\Gamma(x;\alpha n+\beta)q^n}{\Gamma(\alpha n+\beta)}.$$

5   Numerical Analysis of Scale Functions

5.1   Introduction

The methods presented in the previous chapters for producing closed form expressions for scale functions are generous in the number of examples they provide, but also have their limitations. This is not surprising, since the scale function is defined via its Laplace transform, and in most cases it is not possible to find an explicit expression for the inverse of a Laplace transform. Our main objective in this chapter is to present several numerical methods which allow one to compute scale functions for a general spectrally negative Lévy process. As we will see, these computations can be done quite easily and efficiently, however there are a few tricks that one should be aware of. Our second goal is to discuss two very special families of Lévy processes, i.e. processes with jumps of rational transform and Meromorphic processes, for which the scale function can be computed essentially in closed form.

The problem of numerical evaluation of the scale function and other related quantities has received some attention in the literature. In particular, Rogers [89] computes the distribution of the first passage time for spectrally negative Lévy processes by inverting the two-dimensional Laplace transform. The main tool is the discretization of the Bromwich integral and the application of Euler summation in order to improve convergence. We will describe the one dimensional version of this method in Sect. 5.3.2. Surya [99] presents an algorithm for evaluating the scale function using exponential dampening followed by Laplace inversion; the latter performed in a similar way as Rogers [89]. In a recent paper Veillette and Taqqu [103] compute the distribution of the first passage time for subordinators using two techniques. These are the discretization of the Bromwich integral and Post-Widder formula coupled with Richardson extrapolation. Albrecher et al. [6] develop algorithms to compute ruin probabilities for completely monotone claim distributions, which is equivalent to computing the scale function $W(x)$ due to relation (22).

In this section we will present some general ideas related to numerical evaluation of the scale function. We adopt the same notation as in the previous chapters. However, in order to avoid excessive technical details, we shall impose the following condition throughout.

Assumption 1. The Lévy measure Π has at most a finite number of atoms when X has paths of bounded variation.

Recall that the scale function $W^{(q)}(x)$ is defined by the Laplace transform identity

$$\int_0^\infty e^{-zx} W^{(q)}(x) dx = \frac{1}{\psi(z) - q}, \quad \mathrm{Re}(z) > \Phi(q). \tag{92}$$

We know from Lemma 2.4 and Corollary 2.5 that $W^{(q)\prime}(x)$ exists and is continuous everywhere except when X has paths of bounded variation, in which case the derivative does not exist at any point x such that an atom of the Lévy measure occurs at $-x$. Since we have assumed that there exists just a finite number of these points, we can apply standard results (such as Theorem 2.2 in [33]) and conclude that $W^{(q)}(x)$ can be expressed via the Bromwich integral

$$W^{(q)}(x) = \frac{e^{cx}}{2\pi} \int_{\mathbb{R}} \frac{e^{iux}}{\psi(c + iu) - q} du, \quad x \in \mathbb{R}. \tag{93}$$

where c is an arbitrary constant satisfying $c > \Phi(q)$.

In principle one could use (93) as a starting point for numerical Laplace inversion to give $W^{(q)}(x)$. However, this would not be a good approach from the numerical point of view. The problem here is the exponential factor e^{cx}, which can be very large and would amplify the errors present in the numerical evaluation of the integral. Ideally we would like to choose c to be a small positive number, but this is not possible due to the restriction $c > \Phi(q)$. Therefore this method would be reasonable from the numerical point of view only when $q = 0$ and $\Phi(0) = 0$, in which case the function $W^{(q)}(x)$ does not increase exponentially fast. Indeed in such cases there is at most linear growth. This follows on account of the fact that when $q = \Phi(0) = 0$ the underlying Lévy process does not drift to $-\infty$ and W is the renewal measure of the descending ladder height process; recall the discussion preceding (39). Thanks to (39), this in turn implies that $W(x)$ is a renewal function and hence grows at most linearly as x tends to infinity.

In all other cases we would have to modify (93) in order to remove the exponential growth of $W^{(q)}(x)$. It turns out that this can be done quite easily with the help of the density of the potential measure of the dual process $\hat{X} = -X$, defined as

$$\int_0^\infty e^{-qt} \mathbb{P}(\hat{X}_t \in dx)dt = \hat{u}^{(q)}(x)dx. \tag{94}$$

We know from Theorem 2.7 (iv), this function satisfies

$$W^{(q)}(x) = \frac{e^{\Phi(q)x}}{\psi'(\Phi(q))} - \hat{u}^{(q)}(x), \quad x \geq 0, \tag{95}$$

therefore $\hat{u}^{(q)}(x)$ is continuous on $(0, \infty)$ and has finite left and right derivatives at every point $x > 0$. Theorems 3.10–3.12 give us more information on the relation between the Lévy measure of X and the smoothness properties of $\hat{u}^{(q)}(x)$. As we see from identity (95), the problem of computing the scale function $W^{(q)}(x)$ is equivalent to that of computing $\hat{u}^{(q)}(x)$. Dealing with the latter turns out to be an easier problem from the numerical point of view, providing that $q \geq 0$ and $\psi'(0+) < 0$ when $q = 0$, since in this case $u^{(q)}$ is bounded. To see why this is the case, note from the earlier representation of $W^{(q)}$ in (53) together with (17), we have that

$$\hat{u}^{(q)}(x) = \frac{e^{\Phi(q)x}}{\psi'(\Phi(q))} - W^{(q)}(x)$$

$$= \frac{e^{\Phi(q)x}}{\psi'(\Phi(q))} - e^{\Phi(q)x} \frac{1}{\psi'_{\Phi(q)}(0+)} P_x^{\Phi(q)}(\underline{X}_\infty \geq 0)$$

$$= \frac{e^{\Phi(q)x}}{\psi'(\Phi(q))} P_x^{\Phi(q)}(\underline{X}_\infty < 0)$$

$$= \frac{e^{\Phi(q)x}}{\psi'(\Phi(q))} P_x^{\Phi(q)}(\tau_0^- < \infty)$$

$$= \frac{1}{\psi'(\Phi(q))} E_x \left(e^{\Phi(q)X_{\tau_0^-} - q\tau_0^-} \mathbf{1}_{\{\tau_0^- < \infty\}} \right)$$

$$< \frac{1}{\psi'(\Phi(q))},$$

thereby showing boundedness. Note that similar computations to the above can be found in Takács [100] and Bingham [21].

We need to characterize the Laplace transform of $\hat{u}^{(q)}(x)$. The proof of the next proposition follows easily from (95).

Proposition 5.1. *Assume that $\Phi(q) > 0$. Then for $\mathrm{Re}(z) > 0$*

$$\int_0^\infty e^{-zx} \hat{u}^{(q)}(x)\mathrm{d}x = F^{(q)}(z), \tag{96}$$

where

$$F^{(q)}(z) = \frac{1}{\psi'(\Phi(q))(z - \Phi(q))} - \frac{1}{\psi(z) - q}. \tag{97}$$

Let us summarize our approach to computing the scale functions. When $q = 0$ and $\Phi(0) = 0$ we will work with the scale function itself and will use (93) as the starting point for our computations. As earlier noted, in this case $W^{(q)}(x) = W(x)$ grows at worst linearly fast as $x \to +\infty$ and we do not need to worry about amplifications of numerical errors. If $q \geq 0$ and $\psi'(0+) < 0$ when $q = 0$ the scale function grows exponentially fast as $x \to +\infty$, thus it is better to work with the density of the potential measure $\hat{u}^{(q)}(x)$, and then to recover the scale function via relation (95). For convenience however we shall restrict ourselves to the latter case with the following blanket assumption.

Assumption 2. We have $q \geq 0$ and $\psi'(0+) < 0$ when $q = 0$.

The analysis for the case $q = 0$ and $\psi'(0+) \geq 0$, where we work directly with the scale function, is no different.

As we will see later, our algorithms will require the evaluation of $F^{(q)}(z)$ for values of z in the half-plane $\mathrm{Re}(z) > 0$. From (97) we find that $F^{(q)}(z)$ has a removable singularity at $z = \Phi(q)$. It is not advisable to compute $F^{(q)}(z)$ via (97) when z is close to $\Phi(q)$, as this procedure would involve subtracting two large numbers, which would cause the loss of accuracy. There are two solutions to this problem. One should either make sure that z is never too close to $\Phi(q)$ or alternatively one should use the following asymptotic expression for $F^{(q)}(z)$.

Proposition 5.2. *Define $a_n = \psi^{(n)}(\Phi(q))$ where $\psi^{(n)}$ is the n-th derivative of ψ. Then as $z \to \Phi(q)$*

$$F^{(q)}(z) = \frac{1}{2}\frac{a_2}{a_1^2} + \left[\frac{1}{6}\frac{a_3}{a_1^2} - \frac{1}{4}\frac{a_2^2}{a_1^3} \right](z - \Phi(q)) + O((z - \Phi(q))^2). \tag{98}$$

Proof. The proof follows easily by writing down the Taylor expansion of the right-hand side in (97) centered at $z = \Phi(q)$.

As we have seen in Sect. 2.5 and in many other instances, virtually all fluctuation identities for spectrally negative Lévy processes can be expressed in terms of the following three objects: the scale function $W^{(q)}(x)$, its derivative $W^{(q)\prime}(x)$ and a function $Z^{(q)}(x)$ defined by (41), which is essentially an indefinite integral of the scale function. Therefore, in addition to the scale function itself, it is also important to be able to compute its derivative and indefinite integral. It turns out that computation of all these quantities can be done in exactly the same way. Again, the first step is to remove the exponential growth as $x \to +\infty$ and express everything in terms of the potential density $\hat{u}^{(q)}(x)$ as follows

$$W^{(q)\prime}(x) = \frac{\Phi(q)e^{\Phi(q)x}}{\psi'(\Phi(q))} - \hat{u}^{(q)\prime}(x), \quad x \geq 0, \tag{99}$$

and

$$Z^{(q)}(x) = 1 + \frac{q}{\Phi(q)} \frac{e^{\Phi(q)x} - 1}{\psi'(\Phi(q))} - qv^{(q)}(x), \tag{100}$$

where we have defined

$$v^{(q)}(x) = \int_0^x \hat{u}^{(q)}(y)dy. \tag{101}$$

We see that the problem of computing $W^{(q)\prime}(x)$ and $Z^{(q)}(x)$ is equivalent to the problem of computing $\hat{u}^{(q)\prime}(x)$ and $v^{(q)}(x)$. The following result is an analogue of Proposition 5.1 and it gives us Laplace transforms of $\hat{u}^{(q)\prime}(x)$ and $v^{(q)}(x)$.

Proposition 5.3. *Assume that $q \geq 0$ and $\psi'(0+) < 0$ when $q = 0$. Then for* $\mathrm{Re}(z) > 0$

$$\int_0^\infty \hat{u}^{(q)\prime}(x)e^{-zx}dx = zF^{(q)}(z) + W^{(q)}(0^+) - \frac{1}{\psi'(\Phi(q))}, \tag{102}$$

and

$$\int_0^\infty v^{(q)}(x)e^{-zx}dx = \frac{F^{(q)}(z)}{z}. \tag{103}$$

Proof. The proof follows from Proposition 5.1 and integration by parts.

Now we have reduced all three problems to an equivalent "standardized" form. Equations (95), (99), (100) and Propositions (5.1) and (5.3) show that the problem of computing of $W^{(q)}(x)$, $W^{(q)\prime}(x)$ and $Z^{(q)}(x)$ is equivalent to a standard Laplace inversion problem. One has an explicit expression for the function $g(z)$, which is analytic in the half-plane $\mathrm{Re}(z) > 0$ and is equal to the Laplace transform of $f(x)$

$$\int_0^\infty e^{-zx}f(x)dx = g(z), \quad \mathrm{Re}(z) > 0, \tag{104}$$

and one wants to compute the function $f(x)$ for several values of $x > 0$. From now on we will concentrate on solving this problem.

This Laplace inversion problem has been well-studied and it has generated an enormous amount of literature. We will just mention here an excellent textbook by Cohen [33], very helpful reviews and research articles by Abate, Choudhury, Whitt and Valko [1,3,4] and works of Filon [41], Bailey and Swarztrauber [12] and Iserles [50].

This chapter is organized as follows. In Sects. 5.2 and 5.3 we present four general numerical methods for computing $f(x)$. The first approach starts with the Bromwich integral and is based on Filon's method coupled with fractional discrete Fourier transform and Fast Fourier Transform techniques. The next three methods, i.e. Gaver-Stehfest, Euler and Talbot algorithms, also start with the Bromwich integral, but the discretization of this integral is done in a different way, and in every case (except for Talbot method) some acceleration procedure is applied. These three methods typically require multi-precision arithmetic. In Sects. 5.4 and 5.5 we present two families of processes for which the scale function can be computed explicitly, as a finite sum or an infinite series, which involve exponential functions, derivative of the Laplace exponent $\psi(z)$ and the solutions to equation $\psi(z) = q$. In this case computing the scale function can be done in an extremely efficient and accurate way, and later we will use these processes as benchmarks to test the performance of the four general methods. In Sect. 5.6 we discuss the results of several numerical experiments and in Sect. 5.7 we present our conclusions and provide some recommendations on how to choose the right numerical algorithm.

5.2 Filon's Method and Fractional Fast Fourier Transform

The starting point for Filon's method is an expression for $f(x)$ which identifies it as a Bromwich integral. That is,

$$f(x) = \frac{1}{2\pi i} \int_{c+i\mathbb{R}} g(z)e^{zx}dz, \quad x > 0, \tag{105}$$

where c is an arbitrary positive constant. Since we are only interested in $f(x)$ for positive values of x, Eq. (105) can be written in terms of the cosine transform as follows,

$$f(x) = \frac{2e^{cx}}{\pi} \int_0^\infty \text{Re}\left[g(c + iu)\right] \cos(ux)du, \quad x \geq 0. \tag{106}$$

Now our plan is to compute the above integral numerically.

The integral in (106) is an *oscillatory integral*, which means that the integrand is an oscillating function. Evaluating oscillatory integrals is not as straightforward as it may seem, and one should be careful to choose the right numerical method. For example, in (106), the cosine function can cause problems. When x is large it oscillates rapidly with the period $2\pi/x$. Therefore if we simply discretize this integral using the trapezoid or Simpson's rule, we have to make sure that the spacing of the discretization h satisfies $h \ll 2\pi/x$. When x is large, this restriction forces us to take h to be a very small number, therefore we need a huge number of discretization points and the algorithm becomes slow and inefficient. The main benefit of Filon's method is that it helps one to avoid all these problems.

Filon's method applies to computing oscillatory integrals over a finite interval $[a, b]$ of the form

$$\mathscr{F}_c G(x) = \int_a^b G(u) \cos(ux) du. \tag{107}$$

Here $\mathscr{F}_c G$ stands for *cosine transform* of the function G. In order to describe the intuition behind Filon's method, let us revisit Simpson's rule. It is well known that Simpson's rule for the integral in (107) can be obtained in the following way. We discretize the interval $[a, b]$, approximate function $u \mapsto G(u) \cos(ux)$ by the second order Lagrange interpolating polynomials in the u-variable on each subinterval, and then integrate this approximation over $[a, b]$. Filon's method goes along the same steps, except for one crucial difference. The function $G(u)$ is approximated by second order Lagrange interpolating polynomials, which is then multiplied by $\cos(ux)$ and integrated over $[a, b]$. Due to the fact that the product of trigonometric functions and polynomials can be integrated explicitly, we still have an explicit formula. In doing this we separate the effects of oscillation and approximation. The approximation for the function $G(u)$ is usually quite smooth and does not change very fast. Then any effect of oscillation disappears since we compute the integral against $\cos(ux)$ explicitly.

Let us describe this algorithm in full detail. Our main references are [41, 43, 50]. We take N to be an integer number and define $h = (b - a)/(2N)$ and $u_n = a + nh$, $0 \le n \le 2N$. Then we denote $g_n = G(u_n)$, $0 \le n \le 2N$ and introduce the vectors $\mathbf{u} = [u_0, u_1, \ldots, u_{2N}]$ and $\mathbf{g} = [g_0, g_1, \ldots, g_{2N}]$. For $k \in \{1, 2\}$ we define

$$\mathscr{C}_k(\mathbf{g}, \mathbf{u}, x) = \sum_{n=0}^{N-1} g_{2n+k} \cos(x u_{2n+k}). \tag{108}$$

Next, on each subinterval $[u_{2n}, u_{2n+2}]$, $0 \le n < N$ we approximate $G(u)$ by a Lagrange polynomial of degree two, which gives us a composite approximation of the following form

$$G(u; N) = \sum_{n=0}^{N-1} \mathbf{1}_{\{u_{2n} \le u < u_{2n+2}\}} \left[g_{2n+1} + \frac{1}{2h}(g_{2n+2} - g_{2n})(u - u_{2n+1}) \right.$$

$$\left. + \frac{1}{2h^2}(g_{2n+2} - 2g_{2n+1} + g_{2n})(u - u_{2n+1})^2 \right].$$

Multiplying this function by $\cos(ux)$, integrating over the interval $[a, b]$ and simplifying the resulting expression we obtain the final expression for Filon's method,

$$
\mathscr{F}_c G(x; N) = \int_a^b G(u; N) \cos(ux) du \tag{109}
$$

$$
= hA(hx)(G(b)\sin(bx) - G(a)\sin(ax))
$$

$$
+ hB(hx)\left[\mathscr{C}_2(\mathbf{g}, \mathbf{u}, x) - \frac{1}{2}(G(b)\cos(bx) - G(a)\cos(ax))\right]
$$

$$
+ hC(hx)\mathscr{C}_1(\mathbf{g}, \mathbf{u}, x),
$$

where

$$
A(\theta) = \frac{1}{\theta} + \frac{\sin(2\theta)}{2\theta^2} - \frac{2\sin(\theta)^2}{\theta^3}, \tag{110}
$$

$$
B(\theta) = 2\left[\frac{1 + \cos(\theta)^2}{\theta^2} - \frac{\sin(2\theta)}{\theta^3}\right], \tag{111}
$$

$$
C(\theta) = 4\left[\frac{\sin(\theta)}{\theta^3} - \frac{\cos(\theta)}{\theta^2}\right]. \tag{112}
$$

Note that by construction, Filon's approximation is exact for polynomials of degree two or less. It is also known that the error of Filon's approximation is $O(h^3)$, provided that $G^{(3)}(u)$ is continuous, see [43].

We see that in order to evaluate Filon's approximation (109) we have to compute two finite sums $\mathscr{C}_k(\mathbf{g}, \mathbf{u}, x)$ defined by (108). If we want to compute $\mathscr{F}_c G(x; N)$ for just a single value of x then this obviously requires $O(N)$ operations. By the same reasoning, if we want to compute $\mathscr{F}_c G(x; N)$ for N equally spaced points $\mathbf{x} = [x_0, x_1, \ldots, x_{N-1}]$, then we would have to perform $O(N^2)$ operations. However, the special structure of these sums makes it possible to perform the latter computations in just $O(N \ln(N))$ computations. The main idea is to use *fast Fourier transform* (FFT). This works as follows. Assume that we want to compute $\mathscr{F}_c G(x; N)$ for $N_x = N$ values $x_m = x_0 + m\delta_x$, $0 \leq m < N$. We rewrite the finite sums in (109) in the following form

$$
\mathscr{C}_k(\mathbf{g}, \mathbf{u}, x_m) = \text{Re}\left[e^{i(a+kh)x_m} \sum_{n=0}^{N-1} g_{k,n} e^{i(2h\delta_x)nm}\right], \tag{113}
$$

where we have defined $g_{k,n} = g_{2n+k} e^{i2hnx_0}$. Then, assuming that parameters h and δ_x satisfy

$$
h\delta_x = \frac{\pi}{N}, \tag{114}
$$

the expression in the right-hand side of (113) is exactly in the form of the discrete Fourier transform (see [12]), and it is well-known that it can be evaluated *for all* $m = 0, 1, .., N - 1$ in just $O(N \ln(N))$ using the fast Fourier transform technique.

The restriction (114) is quite unpleasant as it does not allow us to choose the spacing between the discretization points in the u-domain in (107) independently of the spacing in the x-domain. However there is an easy solution to this problem, namely the fractional discrete Fourier transform (see [12]), which is defined as a linear transformation, which maps a vector $\mathbf{v} = [v_0, v_1, \ldots, v_{N-1}]$ into a vector $\mathbf{V} = [V_0, V_1, \ldots, V_{N-1}]$

$$V_m = \sum_{n=0}^{N-1} v_n e^{i\alpha n m}, \quad m = 0, 1, \ldots, N - 1. \tag{115}$$

It turns out that for any $\alpha \in \mathbb{R}$ one can still compute the values of V_m for $m = 0, 1, .., N - 1$ in just $O(N \ln(N))$ operations, see [12] for all the details.

Let us summarize the main steps of algorithm. First, choose a small number $c > 0$, so that the factor e^{cx} is not too large for the values of x that interest us. Then set $a = 0$ and choose the value of the cutoff, i.e. a large number $b > 0$ such that the integral

$$\int_b^\infty \text{Re}\,[g(c + iu)] \cos(ux) du$$

is sufficiently small. Then, choose the number of discretization points N in the u-domain and define $u_n = nb/(2N), 0 \le n \le 2N$ and $g_n = \text{Re}\,[g(c + iu_n)]$. Choose δ_x and x_0 and compute the approximation (109) using the fractional Fast Fourier Transform. This gives us N values of $f(x_0 + m\delta_x)$ for $0 \le m < N$, with the error bound $O(N^{-3})$, at the computational cost of $O(N \ln(N))$ operations.

There is another trick that might be very useful when implementing Filon's method. In many examples the integrand $\text{Re}\,[g(c + iu)]$ in (106) changes quite rapidly when u is small while it changes slowly for large values of u. This means that we would have better precision and would need fewer discretization points if we were able to place more of them near $u = 0$ and fewer of them for large u. This can be easily achieved by dividing the domain of integration

$$\int_0^b \text{Re}\,[g(c + iu)] \cos(ux) du = \sum_{j=0}^{n-1} \int_{b_j}^{b_{j+1}} \text{Re}\,[g(c + iu)] \cos(ux) du, \tag{116}$$

where $0 = b_0 < b_1 < \cdots < b_n = b$. Each integral over $[b_j, b_{j+1}]$ can be evaluated using Filon's method to produce results on the same grid of the x-variable. Choosing b_j so that the spacing $b_{j+1} - b_j$ increases allows us to concentrate more points where they are needed (near $u = 0$) and fewer points in the regions far away from $u = 0$.

5.3   Methods Requiring Multi-precision Arithmetic

The methods presented in this section can give excellent performance, but the price that one has to pay is that they all require multi-precision arithmetic. Our main references for this section are [1,3,4,33]. These methods are grouped together since they all give a similar expression for the approximation to $f(x)$

$$f(x) \sim \frac{1}{x} \sum_{n=0}^{M} a_n g\left(\frac{b_n}{x}\right),$$

where the coefficients a_n and b_n depend only on M and do not depend on functions f and g.

5.3.1   The Gaver-Stehfest Algorithm

The first method that we discuss is the Gaver-Stehfest algorithm, see [3] and Sect. 7.2 in [33]. This algorithm is based on the Gaver's approximation [44], which can be considered as a discrete analogue of the Post-Widder formula (see Sect. 2.3 in [33])

$$f(x) = \lim_{k \to \infty} \frac{(-1)^k}{k!} \left(\frac{k}{x}\right)^{k+1} g^{(k)}\left(\frac{k}{t}\right).$$

It turns out that both Gaver's and Post-Widder formulas have a very slow convergence rate, therefore one has to apply some acceleration algorithm, such as Salzer transformation for the Gaver's formula, which was proposed by Stehfest [97], or Richardson extrapolation for Post-Widder formula which was used by Veillette and Taqqu [103].

We refer to [3] for all the details and background on the Gaver-Stehfest algorithm, here we just present the final expression. The function $f(x)$ is approximated by $f^{GS}(x; M)$, which depends on a single integer parameter M and is defined as

$$f^{GS}(x; M) = \frac{\ln(2)}{x} \sum_{n=1}^{2M} a_n g\left(n \ln(2) x^{-1}\right), \tag{117}$$

where the coefficients a_n are given by

$$a_n = (-1)^{M+n} \sum_{j=[(n+1)/2]}^{n \wedge M} \frac{j^{M+1}}{M!} \binom{M}{j} \binom{2j}{j} \binom{j}{n-j}.$$

Note that the coefficients a_n are defined as finite sums of possibly very large numbers, and that these coefficients have alternating signs. This means that we will lose accuracy in (117) due to subtracting very large numbers, thus we have to use multi-precision arithmetic to avoid this problem. Abate and Valko [1] (see also [3]) recommend the following "rule of thumb". If we want j significant digits in our approximation, we should set $M = \lceil 1.1j \rceil$ (the least integer greater than or equal to $1.1j$) and set the system precision at $\lceil 2.2M \rceil$. It is also useful to check the accuracy of computation of a_n using the fact that

$$\sum_{n=1}^{2M} a_n = 0.$$

We see that Gaver-Stehfest algorithm should have efficiency around $0.9/2.2 \sim 0.4$, which is the ratio of the system precision to the number of significant digits produced by the approximation. While algorithms presented below all have higher efficiency, the Gaver-Stehfest algorithm has a big advantage that it does not require the use of complex numbers. This can improve performance, since it is faster to perform computations with real numbers than with complex numbers. This feature is also useful, because not every multi-precision software has a built-in support for complex numbers.

5.3.2 The Euler Algorithm

The next two algorithms are based on the Bromwich integral representation, which follows from (105) after the change of variables of integration $u \mapsto v/x$

$$f(x) = \frac{e^c}{2\pi x} \int_{\mathbb{R}} g\left(\frac{c+iv}{x}\right) e^{iv} dv, \quad x \in \mathbb{R}. \tag{118}$$

Note that we have also changed $c \mapsto c/x$. The main idea behind Euler algorithm is to approximate the integral in (118) by a trapezoid rule and then apply Euler acceleration method to improve the convergence rate. Again, all the details can be found in [4] and [3]. We present here only the final form of this approximation

$$f^E(x; M) = \frac{10^{\frac{M}{3}}}{x} \sum_{n=0}^{2M} (-1)^n a_n \text{Re}\left[g\left(\left(\ln\left(10^{\frac{M}{3}}\right) + \pi in\right) x^{-1}\right)\right], \tag{119}$$

where the coefficients a_n are defined as

$$a_0 = \frac{1}{2}, \quad a_n = 1, \text{ for } 1 \leq n \leq M, \quad a_{2M} = 2^{-M},$$

$$a_{2M-k} = a_{2M-k+1} + 2^{-M} \binom{M}{n}, \text{ for } 1 \leq n < M.$$

Note that while coefficients a_n are real, formula (119) still requires evaluation of $g(z)$ for complex values of z.

In the case of Euler algorithm, Abate and Whitt [3] recommend to set $M =$ $\lceil 1.7j \rceil$ if j significant digits of are required, and then set the system precision at M. Coefficients a_n can be precomputed, and the accuracy can be verified via condition

$$\sum_{n=0}^{2M} (-1)^n a_n = 0.$$

We see that the efficiency of the Euler algorithm should be around $1/1.7 \sim 0.6$.

5.3.3   The Fixed Talbot Algorithm

The Talbot algorithm [101] also starts with the integral representation (118), but then the contour of integration is transformed so that $\text{Re}(z) \to -\infty$ on this contour. Note that $\text{Re}(z)$ is constant on the contour of integration in the Bromwich integral (118). This transformation of the contour of integration has a great benefit in that the integrand $g(z) \exp(zx)$ converges to zero much faster. On the negative side, this method only works when (1) $g(z)$ can be analytically continued into the domain $|\arg(z)| < \pi$, and (2) $g(z)$ has no singularities far away from the negative half-line. In our case these two conditions are satisfied for processes whose jumps have completely monotone density, as in this case all of the singularities of $F^{(q)}(z)$ lie on the negative half-line. Meromorphic Lévy processes, presented in Sect. 5.5, exhibit this property for example. It is also worthy of note for future reference that such processes are dense in the class of processes with a completely monotone jump density. As we will see later, the Talbot method performs excellently for processes with completely monotone jumps.

Again, we refer to [1, 3] and [33] for all the details, and present here only the final form of the approximation

$$f^T(x; M) = \frac{1}{x} \sum_{n=0}^{M-1} \text{Im} \left[a_n g \left(b_n x^{-1} \right) \right], \tag{120}$$

where

$$b_0 = \frac{2M}{5}, \quad b_n = \frac{2\pi n}{5} \left(\cot \left(\frac{\pi n}{M} \right) + i \right), \quad \text{for } 1 \le n < M,$$

and

$$a_0 = \frac{i}{5} e^{b_0}, \quad a_n = \left(b_n - \frac{5}{2M} |b_n|^2 \right) \frac{e^{b_n}}{n\pi}, \quad \text{for } 1 \le n < M.$$

For this algorithm, Abate and Valko [1] (see also [3]) recommend using the same precision parameters as for the Euler algorithm: one should set $M = \lceil 1.7j \rceil$ if j significant digits of are required, and then set the system precision at M. The efficiency of this algorithm should also be close to 0.6.

5.4 Processes with Jumps of Rational Transform

It is well known that the Wiener–Hopf factorization and many related fluctuation identities can be obtained in closed form for processes with jumps of rational transform. The Wiener–Hopf factorization for the two-sided processes having positive and/or negative phase-type jumps (a subclass of jumps of rational transform) was studied by Mordecki [79], Asmussen et al. [8] and Pistorius [84], while Wiener–Hopf factorization for a more general class of processes having positive jumps of rational transform was obtained by Lewis and Mordecki [74]. The reader is also referred to older work of Borovkov [24] for related ideas on the Wiener–Hopf factorization of random walks. Expressions for the scale functions for spectrally negative processes with jumps of rational transform follow implicitly from these papers, their explicit form was obtained in a recent paper by Egami and Yamazaki [39].

In order to specify these processes, let us define the density of the Lévy measure as follows

$$\pi(x) = 1_{\{x<0\}} \sum_{j=1}^{m} a_j |x|^{m_j-1} e^{\rho_j x}, \tag{121}$$

where $m_j \in \mathbb{N}$ and $\operatorname{Re}(\rho_j) > 0$. It is easy to see that the Laplace exponent is given by

$$\psi(z) = \frac{\sigma^2}{2} z^2 + \mu z + \sum_{j=1}^{m} a_j (m_j - 1)! \left[(\rho_j + z)^{-m_j} - \rho_j^{-m_j} \right] \tag{122}$$

and that $\psi(z)$ is a rational function

$$\psi(z) = \frac{P(z)}{Q(z)}, \tag{123}$$

where $\deg(Q) = M = \sum_{j=1}^{m} m_j$ and $\deg(P) = N$, where $N = M + 2$ (resp. $M + 1$) if $\sigma > 0$ (resp. $\sigma = 0$). The next proposition gives us valuable information about solutions of the equation $\psi(z) = q$.

Proposition 5.4.

(i) *For $q > 0$ or $q = 0$ and $\psi'(0) < 0$ equation $\psi(z) = q$ has one solution $z = \Phi(q)$ in the half-plane $\operatorname{Re}(z) > 0$ and $N - 1$ solutions in the half-plane $\operatorname{Re}(z) < 0$.*

(ii) *For $q = 0$ and $\psi'(0) > 0$ (resp. $\psi'(0) = 0$) equation $\psi(z) = q$ has a solution $z = 0$ of multiplicity one (resp. two) and $N - 1$ (resp. $N - 2$) solutions in the half-plane $\operatorname{Re}(z) < 0$.*

(iii) *There exist at most $M + N - 1$ complex numbers q such that the equation $\psi(z) = q$ has solutions of multiplicity greater than one.*

Proof. The proof of (i) and (ii) is trivial and follows from (123) and the general theory of scale functions. Let us prove (iii). Assume that the equation $\psi(z) = q$ has a solution $z = z_0$ of multiplicity greater than one, therefore $\psi'(z_0) = 0$. Using (123) we find that $\psi'(z_0) = 0$ implies $P'(z_0)Q(z_0) - P(z_0)Q(z_0) = 0$. The polynomial $H(z) = P'(z)Q(z) - P(z)Q(z)$ has degree $M + N - 1$, thus there exist at most $M + N - 1$ distinct points z_k for which $\psi'(z_k) = 0$, which implies that there exist at most $M + N - 1$ values q, given by $q_k = \psi(z_k)$, for which the equation $\psi(z) = q$ has solutions of multiplicity greater than one.

The statement in Proposition 5.4 (iii) is quite important for numerical calculations. While it's proof is quite elementary, we were not able to locate this result in the existing literature. The implication of this result is that for a generic Lévy measure defined by (121) and general $q \geq 0$ it is extremely unlikely that equation $\psi(z) = q$ has solutions of multiplicity greater than one. So for all practical purposes (unless we are dealing with the case $\psi'(0) = q = 0$) we can assume that all the solutions of $\psi(z) = q$ have multiplicity equal to one, as we will see later this will considerably simplify all the formulas.

Computing the scale function, or equivalently, the potential density $\hat{u}^{(q)}(x)$ is a trivial task for this class of processes. Let us consider the general case, and assume that equation $\psi(z) = q$ has n distinct solutions $-\zeta_1, -\zeta_2, \ldots, -\zeta_n$ in the half-plane $\mathrm{Re}(z) < 0$, and the multiplicity of $z = -\zeta_j$ is equal to n_j. Due to (123) it is clear that $N - 1 = \sum_{j=1}^{n} n_j$. Rewriting the rational function $1/(\psi(z) - q)$ as partial fractions

$$\frac{1}{\psi(z) - q} = \frac{1}{\psi'(\Phi(q))(z - \Phi(q))} + \sum_{j=1}^{n} \sum_{k=1}^{n_j} \frac{c_{j,k}}{(z + \zeta_j)^k}, \tag{124}$$

and using Proposition 5.1 we can identify $\hat{u}^{(q)}(x)$ as follows

$$\hat{u}^{(q)}(x) = -\sum_{j=1}^{n} e^{-\zeta_j x} \sum_{k=1}^{n_j} \frac{c_{j,k}}{(k-1)!} x^{k-1}, \quad x > 0. \tag{125}$$

Note, that if all the solutions ζ_i have multiplicity one, i.e. $n_j = 1$ and $n + 1 = N = \deg(P)$, then (124) implies that

$$c_{j,1} = \frac{1}{\psi'(-\zeta_j)},$$

and we have a much simpler expression for the potential density

$$\hat{u}^{(q)}(x) = -\sum_{j=1}^{N-1} \frac{e^{-\zeta_j x}}{\psi'(-\zeta_j)}, \quad x \geq 0. \tag{126}$$

We would like to stress again that Proposition 5.4 (iii) tells us that $\hat{u}^{(q)}(x)$ can be computed with the help of (126) for all but a finite number of q, unless we are dealing with the case $\psi'(0) = q = 0$.

5.5 Meromorphic Lévy Processes

The main advantage of processes with jumps of rational transform is that the numerical computations are very simple and straightforward. Everything boils down to solving a polynomial equation $\psi(z) = q$ and performing a partial fraction decomposition of a rational function $1/(\psi(z) - q)$. On the negative side, it is clear that these processes can only have compound Poisson jumps, while in applications it is often necessary to have processes with jumps of infinite activity or even of infinite variation. Meromorphic Lévy processes, which were recently introduced in [65], solve precisely this problem. They allow for much more flexible modeling of the small jump behavior, yet all the computations can be done with the same efficiency as for processes with jumps of rational transform.

In order to define a spectrally negative Meromorphic process X, let us consider the function

$$\pi(x) = 1_{\{x < 0\}} \sum_{j=1}^{\infty} a_j e^{\rho_j x}, \tag{127}$$

where the coefficients a_j and ρ_j are positive and ρ_j increase to $+\infty$ as $j \to +\infty$. It is easily verified using a monotone convergence argument that the convergence of the series

$$\sum_{j \geq 1} \frac{a_j}{\rho_j^3} < \infty$$

is equivalent to the convergence of the integral

$$\int_{-\infty}^{0} x^2 \pi(x) dx < \infty,$$

see [63], thus $\pi(x)$ can be used to define the density of the Lévy measure. Note that by construction $\pi(-x)$ is a completely monotone function.

Using the Lévy–Khintchine formula we find that the Laplace exponent is given by

$$\psi(z) = \frac{1}{2}\sigma^2 z^2 + \mu z + z^2 \sum_{j \geq 1} \frac{a_j}{\rho_j^2(\rho_j + z)}, \quad z \in \mathbb{C}, \tag{128}$$

and we see that $\psi(z)$ is a meromorpic function which has only negative poles at points $z = -\rho_j$. From the general theory we know that for $q \geq 0$ equation $\psi(z) = q$ has a unique solution $z = \Phi(q)$ in the half-plane $\text{Re}(z) > 0$, and we also know that this solution is real. It can be proven (see [63]) that the same is true for equation

$\psi(-z) = q$. For $q \geq 0$ all the solutions $z = \zeta_j$ of $\psi(-z) = q$ in the halfplane $\mathrm{Re}(z) \geq 0$ are real and they satisfy the interlacing property

$$0 \leq \zeta_1 < \rho_1 < \zeta_2 < \rho_2 < \dots. \tag{129}$$

When $q = 0$ and $\psi'(0) = \mathbb{E}[X_1] \leq 0$ we have $\zeta_1 = 0$, otherwise $\zeta_1 > 0$.

The following proposition gives an explicit formula for the potential density, generalizing (126), which is expressed in terms of the roots ζ_j and the first derivative of the Laplace exponent.

Proposition 5.5.

(i) *If X is Meromorphic, then for all $q > 0$ the function $\hat{u}^{(q)}(x)$ is an infinite mixture of exponential functions with positive coefficients*

$$\hat{u}^{(q)}(x) = -\sum_{j=1}^{\infty} \frac{e^{-\zeta_j x}}{\psi'(-\zeta_j)}, \quad x \geq 0. \tag{130}$$

(ii) *If for some $q > 0$ the function $\hat{u}^{(q)}(x)$ is an infinite mixture of exponential functions with positive coefficients, then X is a Meromorphic process.*

The first statement of the above proposition can be proved using Proposition 5.1 and standard analytical techniques. See for example Corollary 2 and Remark 2 in [65] as well as [64]. The second statement can be established using the same technique as in the proof of Theorem 1 in [63]. In both cases we leave all the details to the reader.

Proposition 5.5 shows that we can compute the potential density and the scale function very easily, provided that we know ζ_j and $\psi'(-\zeta_j)$. There seems little hope to find ζ_j explicitly for a general $q \geq 0$ and hence these quantities have to be computed numerically, which in turn requires multiple evaluations of $\psi(z)$. Computing $\psi(z)$ with the help of the partial fraction decomposition (128) is not the best way to do it, as in general the series will converge rather slowly. Therefore it is important to find examples of Meromorphic processes for which $\psi(z)$ can be computed explicitly. Below we present several such examples, the details can be found in [62,63].

(i) θ-process with parameter $\lambda \in \{3/2, 5/2\}$

$$\psi(z) = \frac{1}{2}\sigma^2 z^2 + \mu z + c(-1)^{\lambda-1/2}(\alpha + z/\beta)^{\lambda-1} \coth\left(\pi\sqrt{\alpha + z/\beta}\right)$$
$$- c(-1)^{\lambda-1/2}\alpha^{\lambda-1} \coth\left(\pi\sqrt{\alpha}\right). \tag{131}$$

(ii) β-process with parameter $\lambda \in (1,2) \cup (2,3)$

$$\psi(z) = \frac{1}{2}\sigma^2 z^2 + \mu z + cB(1 + \alpha + z/\beta, 1 - \lambda) - cB(1 + \alpha, 1 - \lambda), \tag{132}$$

where $B(x,y) = \Gamma(x)\Gamma(y)/\Gamma(x+y)$ is the Beta function.

The admissible set of parameters is $\sigma \geq 0$, $\mu \in \mathbb{R}$, $c > 0$, $\alpha > 0$ and $\beta > 0$. The parameters a_j and ρ_j, which define the Lévy measure via (127), are given as follows (see [62, 63]): in the case of θ-process we have

$$a_j = \frac{2}{\pi} c \beta j^{2\lambda - 1}, \quad \rho_j = \beta(\alpha + j^2), \tag{133}$$

and in the case of β-process

$$a_j = c\beta \binom{j + \lambda - 2}{j - 1}, \quad \rho_j = \beta(\alpha + j). \tag{134}$$

In the case of β-process we also have an explicit formula for the density of the Lévy measure,

$$\pi(x) = c\beta \frac{e^{(1+\alpha)\beta x}}{(1 - e^{\beta x})^\lambda}, \quad x < 0.$$

Moreover, it can be shown that the density of the Lévy measure $\pi(x)$ satisfies (up to a multiplicative constant)

$$\pi(x) \sim |x|^{-\lambda}, \quad \text{as } x \to 0^-,$$
$$\pi(x) \sim e^{\beta(1+\alpha)x}, \quad \text{as } x \to -\infty.$$

In particular, we see that the Lévy measure always has exponential tails and the rate of decay of $\pi(x)$ is controlled by parameters α and β. See [62, 63]. The parameter c controls the overall "intensity" of jumps, while λ is responsible for the behavior of small jumps: if $\lambda \in (1, 2)$ (resp. $\lambda \in (2, 3)$) then the jump part of the process is of infinite activity and finite (resp. infinite) variation.

The beta family of processes is more general than the theta family, as it allows for a greater range of the parameter λ, which gives us more flexibility in modeling the behavior of small jumps. However, processes in the theta family have the advantage that the Laplace exponent is given in terms of elementary functions, which helps to implement faster numerical algorithms.

Computing the scale function for Meromorphic processes is very simple. The first step is to compute the values of ζ_j, which are defined as the solutions to $\psi(-z) = q$. The interlacing property (129) gives us left/right bounds for each ζ_j (recall that ρ_j are known explicitly), thus we know that on each interval (ρ_j, ρ_{j+1}) there is a unique solution to $\psi(-z) = q$, and this solution can be found very efficiently using bisection/Newton's method.

There is one additional trick that can greatly reduce the computation time for this first step. We will explain the main idea on the example of a β-process. Formula (134) shows that the spacing between the poles ρ_j is constant and is equal to β. This fact and the behavior of $\psi(z)$ as $\text{Re}(z) \to \infty$ implies that for j large, the difference $\zeta_{j+1} - \zeta_j$ would also be very close to β. Therefore, once we have computed ζ_j for a sufficiently large j, we can use $\zeta_j + \beta$ as a starting point for our search for ζ_{j+1}. In practice, starting Newton's method from this point gives the required accuracy in just one or two iterations. A corresponding strategy can be developed for the θ-processes: one should use the fact that $\sqrt{\rho_j - \alpha\beta}$ have constant spacing equal to $\sqrt{\beta}$.

Once we have precomputed the values of ζ_j, we compute the coefficients $\psi'(-\zeta_j)$, which is an easy task since we have an explicit formula for $\psi'(z)$. Now we can evaluate $\hat{u}^{(q)}(x)$ by truncating the infinite series in (130). Note that the infinite series converges exponentially fast (and much faster in the case of θ-process), so unless x is very small, one really needs just a few terms to have good precision. On the other hand, when x is extremely small, it is better to compute the scale function by using the information about $W^{(q)}(0^+)$ and $W_+^{(q)\prime}(0)$ given in Lemmas 3.1 and 3.2 and applying interpolation.

5.6 Numerical Examples

In this section we present the results of several numerical experiments. Before we start discussing the details of these experiments, let us describe the computing environment. All the code was written in Fortran90 and compiled with the help of Intel® Fortran compiler. For the methods which require multi-precision arithmetic we have the following two options. The first one is the *quad* data type (standard on Fortran90), which uses 128 bits to represent a real number and allows for the precision of approximately 32 decimal digits. Recall that the *double* data type (standard on C/C++/Fortran/Matlab and other programming languages) uses 64 bits and gives approximately 16 decimal digits. The second option is to use MPFUN90 multi-precision library developed by Bailey [11]. This is an excellent library for Fortran90, which is very efficient and easy to use. It allows one to perform computations with precision of thousands of digits, though in our experiments we've never used more than two hundred digits. All the computations were performed on a standard 2008 laptop (Intel Core 2 Duo 2.5 GHz processor and 3 GB of RAM).

In our first numerical experiment we will compare the performance of the four methods described in Sects. 5.2 and 5.3 in the case of a θ-process with $\lambda = 3/2$ (process with jumps of finite variation and infinite activity), whose Laplace exponent is given by (131). We will consider two cases $\sigma = 0$ or $\sigma = 0.25$ and will fix the other parameters as follows

$$\mu = 2, \ c = 1, \ \alpha = 1, \ \beta = 0.5.$$

Moreover, **everywhere in this section we fix** $q = 0.5$.

As the benchmark for this experiment, we compute the values of the scale function $W^{(q)}(x)$ for one hundred values of x, given by $x_i = i/20$, $i = 1$, $2, \ldots, 100$. This part is done using the algorithm described in Sect. 5.6. The infinite series (130) is truncated at $j = 1,000$ and the system precision is set at 200 digits

Table 1 Computing $W^{(q)}(x)$ for the θ-process with $\sigma = 0.25$

Algorithm	Rel. error	Time (seconds)	N_x	M	System precision
Filon ($b = 10^7$)	3.0e-12	0.18	10,000	–	double (64 bits)
Gaver-Stehfest	9.2e-12	0.4	100	20	44 (MPFUN90)
Euler	9.0e-14	0.05	100	20	quad (128 bits)
Talbot	2.1e-13	0.025	100	20	quad (128 bits)
Gaver-Stehfest	2.9e-21	1.5	100	40	88 (MPFUN90)
Euler	2.7e-25	1.5	100	40	40 (MPFUN90)
Talbot	2.4e-25	0.7	100	40	40 (MPFUN90)
Gaver-Stehfest	2.2e-39	6.3	100	80	176 (MPFUN90)
Euler	2.1e-48	5.0	100	80	80 (MPFUN90)
Talbot	3.1e-49	2.5	100	80	80 (MPFUN90)

(using MPFUN90 library). This guarantees that our benchmark is within 1.0e-150 of the exact value.

Next we compute the approximation $\tilde{W}^{(q)}(x)$ using Filon/Gaver-Stehfest/Euler/Talbot algorithms for the same set of values of x. As the measure of accuracy of the approximation we will take the maximum of the relative error

$$\text{relative error} = \max_{1 \leq i \leq 100} \frac{|\tilde{W}^{(q)}(x_i) - W^{(q)}(x_i)|}{W^{(q)}(x_i)}. \tag{135}$$

Let us specify the parameters for the Filon's method, as described in Sect. 5.2. We truncate the integral at $b = 10^7$ (or $b = 10^9$) in (116) and divide the domain of integration into ten sub-intervals, i.e. $n = 10$. Moreover, we set $b_j = b(j/n)^5$ for $j = 1, 2, \ldots, n$ so that there are more discretization points close to $u = 0$. The number of discretization points per each sub-interval $[b_j, b_{j+1}]$ is fixed at $N_x = 10^4$ (or $N_x = 10^5$). Note that Filon's method as described in Sect. 5.2 produces N_x values of $W^{(q)}(x)$. The spacing between the points in x-domain is chosen at $\delta_x = 5/N_x$. In this way the grid $x_m = m\delta_x$ covers the whole interval $[0, 5]$. In order to perform the Fast Fourier Transform we use Intel® MKL library.

The results of this numerical experiment for computing $W^{(q)}(x)$ are presented in Table 1. We see that each of the Gaver-Stehfest/Euler/Talbot algorithms produce very accurate results. When $M = 40$ we obtain more than twenty significant digits and when $M = 80$ we get almost fifty significant digits. The computation time varies with each algorithm and also depends on parameters. Note that when we use the quad date type for Euler/Talbot methods with $M = 20$, the computations are very fast, on the order of one hundredth of a second. When we use MPFUN90 library the computation time increases significantly. This is due to the fact that quad is a native data type in Fortran90 and computations done with an external multi-precision library are slower. Next we see that Talbot algorithm is always about two times faster than Euler method. This is due to the fact that the finite sum in (120) has M terms while Euler method (119) requires summation of $2M + 1$ terms. We would

Table 2 Computing $W^{(q)'}(x)$ for the θ-process with $\sigma = 0$

Algorithm	Rel. error	Time (seconds)	N_x	M	System precision
Filon ($b = 10^9$)	8.0e-5	2.9	10^5	–	double (64 bits)
Gaver-Stehfest	4.6e-12	0.45	100	20	44 (MPFUN90)
Euler	1.7e-13	0.05	100	20	quad (128 bits)
Talbot	1.5e-12	0.023	100	20	quad (128 bits)
Gaver-Stehfest	7.6e-21	1.5	100	40	88 (MPFUN90)
Euler	4.3e-25	1.4	100	40	40 (MPFUN90)
Talbot	2.9e-24	0.71	100	40	40 (MPFUN90)
Gaver-Stehfest	1.7e-38	6.3	100	80	176 (MPFUN90)
Euler	2.8e-48	4.9	100	80	80 (MPFUN90)
Talbot	9.1e-48	2.4	100	80	80 (MPFUN90)

like to stress again that the time values presented in this table are for computing the scale function for N_x different values of x. For example, if one wants to compute the scale function for a *single* value of x using Talbot method with $M = 40$, it would take just *seven milliseconds*.

Filon's method also performs well, computing the scale function for $10,000$ values of x with accuracy of around 12 decimal digits in just 0.18 seconds. As we see from the table, the Talbot algorithm appears to be the most efficient. This is not surprising given the discussion in Sect. 5.3.3 where it was pointed out that this algorithm is well suited for processes with completely monotone jumps (see also results by Albrecher et al. [6] on ruin probabilities for completely monotone claim distributions).

In Table 2 we present the results of computing the first derivative of the scale function. The results are very similar to those presented in Table 1, the major difference now is that Filon's method is not producing a very good accuracy. This happens because the integrand in (106) decays very slowly (see Proposition 5.3), and even though we have increased b and N_x, we still get only five significant digits.

For our second numerical experiment we will consider a process with jumps of rational transform, which are not completely monotone. We define the density of the Lévy measure as follows

$$\pi(x) = \mathbf{1}_{\{x<0\}} e^{-x} (1 + \cos(ax)). \tag{136}$$

Note that this is an example of a process which has jumps of rational transform but whose distribution does not belong to the phase-type class. Recall that any distribution from the phase-type class can be characterized as the time to absorption in a finite state Markov chain with one absorbing state when the chain is started from a particular state. One can easily see that $\pi(-x)$ can not be the density of such an absorption time, since it is equal to zero for $x = (2k + 1)\pi/(a)$. The Laplace exponent of this process can be computed using formula (122)

$$\psi(z) = \frac{1}{2}\sigma^2 z^2 + \mu z + \frac{1}{z+1} + \frac{z+1}{(z+1)^2 + a^2} - 1 - \frac{1}{1+a^2}.$$

Table 3 Computing $W^{(q)}(x)$ for the process with jumps of rational transform

Algorithm	Rel. error	Time (seconds)	N_x	M	System precision
Filon ($b = 10^7$)	1.2e-11	0.14	10 000	–	double (64 bits)
Gaver-Stehfest	1.2e-5	0.06	100	20	44 (MPFUN90)
Euler	9.1e-14	0.048	100	20	quad (128 bits)
Talbot	9.2e-5	0.022	100	20	quad (128 bits)
Gaver-Stehfest	5.5e-10	0.23	100	40	88 (MPFUN90)
Euler	2.6e-25	0.44	100	40	40 (MPFUN90)
Talbot	2.8e-13	0.22	100	40	40 (MPFUN90)
Gaver-Stehfest	1.5e-27	1.2	100	80	176 (MPFUN90)
Euler	2.0e-48	1.2	100	80	80 (MPFUN90)
Talbot	2.9e-49	0.6	100	80	80 (MPFUN90)

We fix the parameters at

$$\sigma = 0.25, \ \mu = 2, \ a = 4.$$

For this set of parameters we have $\Phi(q) \sim 0.37519$ (recall that we have fixed $q = 0.5$ throughout this section) and all the roots of $\psi(-z) = q$ are simple and are located at

$$\zeta_{1,2} \sim 0.97265 \pm 4.0518i, \ \zeta_3 \sim 0.64448, \ \zeta_4 \sim 64.7854$$

Once we have these roots the potential density $\hat{u}^{(q)}(x)$ can be computed from (126) and the scale function using relation (95). The parameters for Filon's method are the same as in our previous experiment with θ-process. The results are presented in Table 3 and are seen to be similar to those of the θ-process, with the exception that the Talbot method is not performing as well for small values of M. This is most likely due to the fact that function $F^{(q)}(z)$ has non-real singularities in the half-plane $\text{Re}(z) < 0$ and, as discussed in Sect. 5.3.3, this may cause problems for the Talbot method. However, when $M = 80$, the Talbot method is performing very well just as it did for the case of the θ-process.

For our final numerical experiment, we consider a spectrally negative Lévy process with the Lévy measure defined by

$$\Pi(dx) = c_1 1_{\{x<0\}} \frac{e^{\lambda_1 x}}{|x|^{1+\alpha_1}} dx + c_2 1_{\{x<0\}} \frac{1}{|x|^{1+\alpha_2}} dx \tag{137}$$

$$+ c_3 \delta_{-a_3}(dx) + c_4 1_{\{x<-a_4\}} e^x dx.$$

Here the admissible range of parameters is $c_i \geq 0$, $\lambda_1 > 0$, $\alpha_1 \in (-\infty, 2) \setminus \{0, 1\}$, $\alpha_2 \in (0, 2) \setminus \{1\}$ and $a_i > 0$. Using the Lévy–Khintchine formula and standard results on tempered stable processes (see Proposition 4.2 in [34]) we find that the Laplace exponent of X can be computed as follows

Table 4 Parameter sets

	σ	μ	c_1	λ_1	α_1	c_2	α_2	c_3	a_3	c_4	a_4
Set 1	0	2	1	1	0.5	1	1.5	0	–	0	–
Set 2	0	2	0.05	1	−5	0.25	1.5	0	–	0	–
Set 3	0.25	2	1	1	0.5	0	–	0	–	1	1
Set 4	0	2	1	1	0.5	0	–	1	1	0	–

$$\psi(z) = \frac{1}{2}\sigma^2 z^2 + \mu z + c_1 \Gamma(-\alpha_1) \left\lfloor (z+\lambda_1)^{\alpha_1} - \lambda_1^{\alpha_1} - z\alpha_1\lambda_1^{\alpha_1-1} \right\rfloor \quad (138)$$

$$+ c_2 \Gamma(-\alpha_2) z^{\alpha_2} + c_3 \left[e^{-a_3 z} - 1 \right] + c_4 \left[\frac{e^{-a_4 z}}{z+1} - 1 \right].$$

We see that the Lévy measure (137) is a mixture of (a) the Lévy measure of a tempered stable process, (b) the Lévy measure of a stable process, (c) an atom at $-a_3$ and (d) an exponential jump distribution shifted by $-a_4$. Table 4 specifies parameter choices for four different sets of Lévy processes which capture the following characteristics.

Set 1: Jumps of infinite variation with a completely monotone Lévy density; no Gaussian component.

Set 2: Jumps of infinite variation which have smooth, but not completely monotone Lévy density; no Gaussian component.

Set 3: Jumps of infinite activity, finite variation and discontinuous Lévy density; non-zero Gaussian component.

Set 4: Jumps of infinite activity, finite variation, the Lévy measure has an atom; no Gaussian component.

In order to compare the performance of the algorithms we need to have a benchmark. In this experiment we don't have any explicit formula for the scale function, thus we have to compute the benchmark numerically. We have chosen to use Filon's method for this purpose as it is quite robust. For each case we have tried different parameters to ensure that we have at least 10–11 digits of accuracy. For example, for the parameter Set 4 we fix $b = 10^8$, $c = 0.1$, divide the interval of integration in (116) into 1,000 subintervals and set $b_j = b(j/n)^2$ for $j = 1, 2, \ldots, n$ and $n = 1,000$. Finally we fix the number of discretization points for each subinterval at $N = N_x = 5 \times 10^6$.

As in our previous numerical experiment, we compute the scale function for one hundred values of x, given by $x_i = i/20$, $i = 1, 2, \ldots, 100$. The relative errors presented below are maximum relative errors over this range of x-values.

First we compute the scale function for the parameter Set 1 and present the results in Table 5. In this case we have a process with completely monotone jumps, and we see that the situation is exactly the same as in our earlier experiment with θ-process. All four methods provide good accuracy, although the Talbot algorithm is faster and gives better accuracy.

Table 5 Computing $W^{(q)}(x)$ for the process with parameter Set 1

Algorithm	Rel. error	Time (seconds)	N_x	M	System precision
Filon ($b = 10^7$)	2.6e-10	0.46	10,000	–	double (64 bits)
Gaver-Stehfest	1.2e-12	1.1	100	20	44 (MPFUN90)
Euler	1.3e-12	0.05	100	20	quad (128 bits)
Talbot	7.3e-13	0.024	100	20	quad (128 bits)

Table 6 Computing $W^{(q)}(x)$ for the process with parameter Set 2

Algorithm	Rel. error	Time (seconds)	N_x	M	System precision
Filon ($b = 10^7$)	8.4e-10	0.47	10,000	–	double (64 bits)
Gaver-Stehfest	3.9e-12	1.1	100	20	44 (MPFUN90)
Euler	3.9e-12	0.05	100	20	quad (128 bits)
Talbot	3.6e-12	0.024	100	20	quad (128 bits)

Table 7 Computing $W^{(q)}(x)$ for the process with parameter Set 3

Algorithm	Rel. error	Time (seconds)	N_x	M	System precision
Filon ($b = 10^7$)	4.6e-11	0.46	10,000	–	double (64 bits)
Gaver-Stehfest	3.3e-5	1.1	100	20	44 (MPFUN90)
Euler	1.4e-6	0.06	100	20	quad (128 bits)
Talbot	5.0e-3	0.028	100	20	quad (128 bits)
Gaver-Stehfest	7.0e-6	4.5	100	40	88 (MPFUN90)
Euler	9.1e-7	4.7	100	40	40 (MPFUN90)
Talbot	1.7e-3	2.3	100	40	40 (MPFUN90)
Gaver-Stehfest	1.1e-6	22	100	80	176 (MPFUN90)
Euler	3.0e-6	18	100	80	80 (MPFUN90)
Talbot	2.9e-4	8.8	100	80	80 (MPFUN90)

In Table 6 we show the results for the parameter Set 2. In this case the jumps have smooth, but not completely monotone density. However, this does not seem to affect the results, and the performance of all four algorithms is similar to the case of parameter Set 1.

In Table 7 we see the first example where the jump density (and therefore the scale function) is non-smooth. In this case we have $W^{(q)}(x) \in C^3(0, \infty)$ (see Theorem 3.11), and we would expect that the discontinuity in the fourth derivative of $W^{(q)}(x)$ shouldn't affect the performance of our numerical methods. Unfortunately, this is not the case. We see that, while Filon's method is still performing well, the Talbot algorithm provides only 2–3 digits of precision. The Gaver-Stehfest and Euler algorithms do a slightly better job and produce 5–6 digits of precision. However, when we increase M we do not see a significant decrease in the error, and, in fact, it is not clear whether these three methods would converge to the right values at all. This behaviour, specifically problems with Laplace inversion when the target function is non-smooth, is quite well-known. See for example the discussion in [4] and Sect. 14 of [2].

Table 8 Computing $W^{(q)}(x)$ for the process with parameter Set 4

Algorithm	Rel. error	Time (seconds)	N_x	M	System precision
Filon ($b = 10^7$)	5.3e-8 (4.6e-11)	5.1	10^5	–	double (64 bits)
Gaver-Stehfest	1.6e-3	1.1	100	20	44 (MPFUN90)
Euler	3.2e-4	0.05	100	20	quad (128 bits)
Talbot	4.6e-3	0.024	100	20	quad (128 bits)
Gaver-Stehfest	8.0e-4	4.5	100	40	88 (MPFUN90)
Euler	3.2e-4	4.6	100	40	40 (MPFUN90)
Talbot	2.4e-3	2.2	100	40	40 (MPFUN90)
Gaver-Stehfest	3.9e-4	22	100	80	176 (MPFUN90)
Euler	8.1e-3	18	100	80	80 (MPFUN90)
Talbot	1.2e-3	8.7	100	80	80 (MPFUN90)

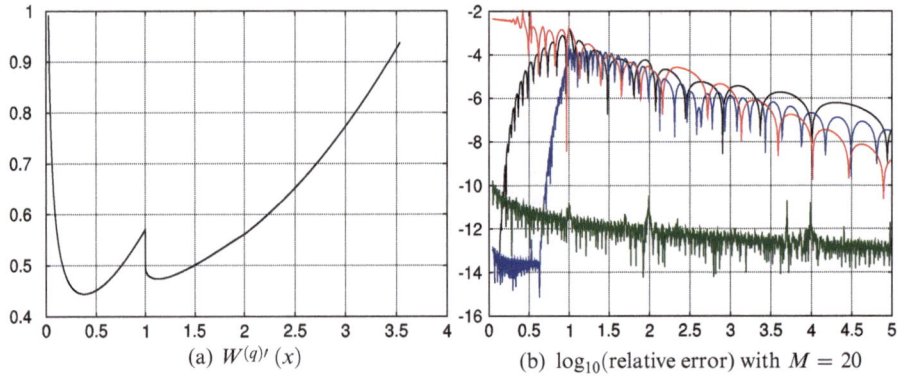

(a) $W^{(q)\prime}(x)$

(b) $\log_{10}$(relative error) with $M = 20$

Fig. 1 Computing $W^{(q)}(x)$ for the process with parameter Set 4. The relative errors produced by the Filon, Gaver-Stehfest, Euler and Talbot methods correspond to the *green*, *black*, *blue* and *red* *curves* respectively

Finally, in Table 8 we present the most extreme case of parameter Set 4. In this case the scale function is not even continuously differentiable, so we should expect to see some interesting behavior of our numerical algorithms. We see that Filon's method is still performing very well. Indeed, the maximum relative error is around 5.3e-8. The error achieves its maximum at $x = 1$, which is not surprising since this is the point where $W^{(q)}(x)$ is not differentiable (see Fig. 1). If we remove this point, then the maximum relative error drops down to 4.6e-11, which is an extremely good degree of accuracy for such non-smooth function. By contrast, the other three algorithms are all struggling to produce even four digits of precision. Again, it is not clear whether these algorithms converge as M increases. In particular we see that the error of the Euler method appears to increase.

We obtain an overview of what is happening in this case by inspecting the graph of $W^{(q)\prime}(x)$ (see Fig. 1). We see that $W^{(q)\prime}(x)$ has a jump at $x = 1$, which is the point where the Lévy measure has an atom. This behavior is expected due to

Corollary 2.5. Then we observe that while $W^{(q)'}(x)$ is continuous at all points $x \neq 1$, the second derivative $W^{(q)''}(x)$ has a jump at $x = 2$. Although there is no theoretical justification for the appearance of such a jump in this text, this observation is consistent with the results on discontinuities in higher derivatives of potential measures of subordinators obtained by Döring and Savov [37].

5.7 Conclusion

In the previous section we have presented results of several numerical experiments, which allow us to compare the performance of four different methods for computing the scale function.

Filon's method has the advantage of being the most robust. We are guaranteed to have an approximation which is within $O(N^{-3})$ of the exact value, provided that we have chosen the cutoff b to be large enough. Another advantage of this method is that it gives not just a single value of $W^{(q)}(x)$, but an array of N values $W^{(q)}(x_i)$ (where the x_i are equally spaced) at the computational cost of just $O(N \ln(N))$ operations. This means that by increasing N we get a better accuracy and, at the same time, more values of $W^{(q)}(x)$, and the computational time would increase almost linearly in N. This is different from the other three methods, where the computational time needed to produce N values of $W^{(q)}(x)$ is $O(NM)$.

At the same time, there are several disadvantages relating to Filon's method. First of all, this algorithm requires quite some effort to program. Secondly, when the integrand decreases very slowly, it may be necessary to take an extremely large value of the cutoff value b. This would have to be compensated by an increase in the number of discretization points N (see Table 2). Finally, unlike the other three methods, which depend on a single parameter M, Filon's method has two important parameters, b and N (as well as b_j and n if one uses (116)), and it is not a priori clear how to choose these parameters. Usually, reasonable values of b and N can be found after some experimentation, which is of course time consuming.

The essential property of the other three methods, i.e the Gaver-Stehfest, Euler and Talbot algorithms, is that all of them require multi-precision arithmetic. If the scale function is smooth and we need just 10–12 digits of precision, then we can use the quad (128 bits) precision, which is available in some programming languages, such as Fortran90. In all other cases one has to use specialized software which is capable of working with high-precision numbers. Whilst very efficient multi-precision arithmetic libraries do exist, such as MPFUN90 by Bailey [11], they are usually much slower than the native double/quad precision of any programming language.

An advantage of the Gaver-Stehfest/Euler/Talbot methods is their simplicity. These algorithms are very easy to program, and they all depend only on just a single integer parameter M.

The performance of the Gaver-Stehfest, Euler and Talbot algorithms depends a lot on the smoothness of the scale function $W^{(q)}(x)$. When $W^{(q)}(x)$ is not sufficiently smooth it might be better to use Filon's method. When the process has completely monotone jumps, Talbot's algorithm is arguably preferable to the other methods as it produces excellent accuracy. Moreover, it is almost twice as fast as its nearest competitor, the Euler algorithm. When the process has a smooth, but not completely monotone, jump density the Euler algorithm is recommendable. One can still use the Talbot algorithm in this case, however, this should be done with a caution as it is not guaranteed to work (see the discussion in Sect. 5.3.3 and the results of our second numerical experiment).

Finally, the Gaver-Stehfest algorithm is somewhat special. It is the only algorithm which does not require the evaluation of $\psi(z)$ at complex values of z. Our results show that this method is not the fastest and not the most precise, but it can still be useful in some situations. For example, this method might be the only available option when the Laplace exponent $\psi(z)$ is given by a very complicated formula which is valid for real values of z and, moreover, it is not clear how to analytically continue it into the complex half-plane $\text{Re}(z) > 0$. The Gaver-Stehfest algorithm can also be helpful when one is using multi-precision software which does not support operations on complex numbers.

References

1. J. Abate, P.P. Valko, Multi-precision Laplace transform inversion. Int. J. Numer. Meth. Eng. **60**, 979–993 (2004)
2. J. Abate, W. Whitt, The Fourier-series method for inverting transforms of probability distributions. Queueing Syst. **10**, 5–88 (1992)
3. J. Abate, W. Whitt, A unified framework for numerically inverting Laplace transforms. INFORMS J. Comput. **18**, 408–421 (2006)
4. J. Abate, G.L. Choudhury, W. Whitt, in *An Introduction to Numerical Transform Inversion and Its Application to Probability Models*, ed. by W. Grassman. Computational Probability (Kluwer, Dordrecht, 1999), pp. 257–323
5. H-J. Albrecher, J-F. Renaud, X. Zhou, A Lévy insurance risk process with tax. J. Appl. Probab. **45**, 363–375 (2008)
6. H-J. Albrecher, F. Avram, D. Kortschak, On the efficient evaluation of ruin probabilities for completely monotone claim distributions. J. Comput. Appl. Math. **233**, 2724–2736 (2010)
7. D. Applebaum, *Lévy Processes and Stochastic Calculus*, 2nd edn. (Cambridge University Press, London, 2009)
8. S. Asmussen, F. Avram, M.R. Pistorius, Russian and American put options under exponential phase-type Lévy models. Stoch. Proc. Appl. **109**, 79–111 (2004)
9. F. Avram, A.E. Kyprianou, M. Pistorius, Exit problems for spectrally negative Lévy processes and applications to (Canadized) Russian options. Ann. Appl. Probab. **14**, 215–238 (2004)
10. F. Avram, Z. Palmowski, M. Pistorius, On the optimal dividend problem for a spectrally negative Lévy process. Ann. Appl. Probab. **17**, 156–180 (2007)
11. D.H. Bailey, A fortran 90-based multiprecision system. ACM Trans. Math. Software **21**, 379–387 (1995)
12. D.H. Bailey, P.N. Swarztrauber, The fractional Fourier transform and applications. SIAM Rev. **33**, 389–404 (1991)
13. C. Berg, G. Forst, in *Potential Theory on Locally Compact Abelian Groups*. Ergebnisse der Mathematik und ihrer Grenzgebiete, Band 87 (Springer, Berlin, 1975)

14. J. Bertoin, An extension of Pitman's Theorem for spectrally positive Lévy processes. Ann. Probab. **20**, 1464–1483 (1992)
15. J. Bertoin, *Lévy Processes* (Cambridge University Press, London, 1996)
16. J. Bertoin, On the first exit time of a completely asymmetric stable process from a finite interval. Bull. Lond. Math. Soc. **28**, 514–520 (1996)
17. J. Bertoin, Exponential decay and ergodicity of completely asymmetric Lévy processes in a finite interval. Ann. Appl. Probab. **7**, 156–169 (1997)
18. K. Bichteler, *Stochastic Integration with Jumps* (Cambridge University Press, London, 2002)
19. E. Biffis, A.E. Kyprianou, A note on scale functions and the time value of ruin for Lévy insurance risk processes. Insur. Math. Econ. **46**, 85–91 (2010)
20. E. Biffis, M. Morales, On an extension of the Gerber-Shiu function to path-dependent penalties. Insur. Math. Econ. **46**, 92–97 (2010)
21. N.H. Bingham, Fluctuation theory in continuous time. Adv. Appl. Probab. **7**, 705–766 (1975)
22. N.H. Bingham, Continuous branching processes and spectral positivity. Stochast. Process. Appl. **4**, 217–242 (1976)
23. L. Bondesson, in *Generalized Gamma Convolutions and Related Classes of Distributions and Densities*. Lecture Notes in Statistics, 76 (Springer, New York, 1992)
24. A.A. Borovkov, *Stochastic Processes in Queueing Theory* (Springer, Berlin, 1976)
25. N.S. Bratiychuk, D.V. Gusak, *Boundary Problem for Processes with Independent Increments* (Naukova Dumka, Kiev, 1991) (in Russian)
26. M.E. Caballero, A. Lambert, G. Uribe Bravo, Proof(s) of the Lamperti representation of continuous-state branching processes. Probab. Surv. **6**, 62–89 (2009)
27. T. Chan, A.E. Kyprianou, M. Savov, Smoothness of scale functions for spectrally negative Lévy processes. Probab. Theor. Relat. Fields **150**, 691–708 (2011)
28. L. Chaumont, Sur certains processus de Lévy conditionés a rester positifs. Stochast. Stochast. Rep. **47** (1994) 1–20 (1994)
29. L. Chaumont, Conditionings and path decompositions for Lévy processes. Stochast. Process. Appl. **64**, 34–59 (1996)
30. L. Chaumont, R.A. Doney, On Lévy processes conditioned to stay positive. Electron. J. Probab. **10**, 948–961 (2005)
31. L. Chaumont, A.E. Kyprianou, J.C. Pardo, Some explicit identities associated with positive self-similar Markov processes. Stochast. Process. Appl. **119**(3), 980–1000 (2009)
32. M. Chazal, A.E. Kyprianou, P. Patie, A transformation for Lévy processes with one-sided jumps and applications. Preprint (2010)
33. A.M. Cohen, *Numerical Methods for Laplace Transform Inversion*, vol. 5 of Numerical Methods and Algorithms (Springer, Berlin, 2007)
34. R. Cont, P. Tankov, *Financial Modeling with Jump Processes* (Chapman & Hall, London, 2004)
35. C. Donati-Martin, M. Yor, Further examples of explicit Krein representations of certain subordinators. *Publ. Res. Inst. Math. Sci.* **43**, 315–328 (2007)
36. R.A. Doney, in *Some Excursion Calculations for Spectrally One-Sided Lévy Processes*. Séminaire de Probabilités, vol. XXXVIII. Lecture Notes in Math., 1857 (Springer, Berlin, 2005), pp. 5–15
37. L. Döring, M. Savov, (Non)differentiability and asymptotics for renewal densities of subordinators. Electron. J. Probab. **16**, 470–503 (2011)
38. P. Dube, F. Guillemin, R.R. Mazumdar, Scale functions of Lévy processes and busy periods of finite-capacity M/GI/1 queues. J. Appl. Probab. **41**, 1145–1156 (2004)
39. M. Egami, K. Yamazaki, On scale functions of spectrally negative Lévy processes with phase-type jumps. arXiv:1005.0064v3 [math.PR] (2010)
40. W. Feller, *An introduction to probability theory and its applications*. Vol II, 2nd edition (Wiley, New York, 1971)
41. L.N.G. Filon, On a quadrature formula for trigonometric integrals. Proc. R. Soc. Edinb. **49**, 38–47 (1928)

42. B. de Finetti, Su un'impostazione alternativa dell teoria collecttiva del rischio. Transactions of the XVth International Congress of Actuaries, vol. 2, pp. 433–443 (1957)
43. L.D. Fosdick, A special case of the Filon quadrature formula. Math. Comp. **22**, 77–81 (1968)
44. D.P. Gaver, Observing stochastic processes and approximate transform inversion. Oper. Res. **14**, 444–459 (1966)
45. H.U. Gerber, E.S.W. Shiu, On the time value of ruin. North Am. Actuarial J. **2**, 48–78 (1998)
46. P.E. Greenwood, J.W. Pitman, Fluctuation identities for Lévy processes and splitting at the maximum. Adv. Appl. Probab. **12** 839–902 (1979)
47. J. Hawkes, On the potential theory of subordinators. Z. W. **33**(2), 113–132 (1975)
48. J. Hawkes, Intersections of Markov random sets. Z. W. **37**(3), 243–251 (1976)
49. F. Hubalek, A.E. Kyprianou, Old and new examples of scale functions for spectrally negative Lévy processes. *Sixth Seminar on Stochastic Analysis, Random Fields and Applications, eds R. Dalang, M. Dozzi, F. Russo*. Progress in Probability, Birkhuser (2010) 119–146.
50. A. Iserles, On the numerical quadrature of highly-oscillating integrals I: Fourier transforms. IMA J. Numer. Anal. **24**, 365–391 (2004)
51. N. Jacobs, *Pseudo Differential Operators and Markov Processes. Vol. I Fourier Analysis and Semigroups* (Imperial College Press, London, 2001)
52. L.F. James, B. Roynette, M. Yor, Generalized Gamma Convolutions, Dirichlet means, Thorin measures, with explicit examples. Probab. Surv. **5**, 346–415 (2008)
53. H. Kesten, in *Hitting Probabilities of Single Points for Processes with Stationary Independent Increments*. Memoirs of the American Mathematical Society, No. 93 (American Mathematical Society, Providence, 1969)
54. J.F.C. Kingman, Markov transition probabilities. II. Completely monotonic functions. Z. W. **9**, 1–9 (1967)
55. R. Knobloch, A.E. Kyprianou, Survival of homogenous fragmentation processes with killing (2011), http://arxiv.org/abs/1104.5078[math.PR]
56. V.S. Korolyuk, Boundary problems for a compound Poisson process. Theor. Probab. Appl. **19**, 1–14 (1974)
57. V.S. Korolyuk, *Boundary Problems for Compound Poisson Processes* (Naukova Dumka, Kiev, 1975) (in Russian)
58. V.S. Korolyuk, On ruin problem for compound Poisson process. Theor. Probab. Appl. **20**, 374–376 (1975)
59. V.S. Korolyuk, Ju.V. Borovskich, *Analytic Problems of the Asymptotic Behaviour of Probability Distributions* (Naukova Dumka, Kiev, 1981) (in Russian)
60. V.S. Korolyuk, V.N. Suprun, V.M. Shurenkov, Method of potential in boundary problems for processes with independent increments and jumps of the same sign. Theor. Probab. Appl. **21**, 243–249 (1976)
61. N. Krell, Multifractal spectra and precise rates of decay in homogeneous fragmentations. Stochast. Proc. Appl. **118**, 897–916 (2008)
62. A. Kuznetsov, Wiener-Hopf factorization and distribution of extrema for a family of Lévy processes. Ann. Appl. Probab. **20**, 1801–1830 (2010)
63. A. Kuznetsov, Wiener-Hopf factorization for a family of Lévy processes related to theta functions. J. Appl. Probab. **47**, 1023–1033 (2010)
64. A. Kuznetsov, M. Morales, Computing the finite-time expected discounted penalty function for a family of Lévy risk processes. Preprint (2010)
65. A. Kuznetsov, A.E. Kyprianou, J.C. Pardo, Meromorphic Lévy processes and their fluctuation identities. Ann. Appl. Probab. **22**, 1101–1135 (2012)
66. A.E. Kyprianou, *Introductory Lectures on Fluctuations of Lévy Processes and Their Applications* (Springer, Berlin, 2006)
67. A.E. Kyprianou, J-C. Pardo, Continuous state branching processes and self-similarity. J. Appl. Probab. **45**, 1140–1160 (2008)
68. A.E. Kyprianou, V. Rivero, Special, conjugate and complete scale functions for spectrally negative Lévy processes. Electron. J. Probab. **13**, 1672–1701 (2008)

69. A.E. Kyprianou, X. Zhou, General tax structures and the Lévy insurance risk model. J. Appl. Probab. **46**, 1146–1156 (2009)
70. A.E. Kyprianou, V. Rivero, R. Song, Convexity and smoothness of scale functions and de Finetti's control problem. J. Theor. Probab. **23**, 547–564 (2010)
71. A.E. Kyprianou, R.L. Loeffen, J.L. Pérez, Optimal control with absolutely continuous strategies for spectrally negative Lévy processes. J. Appl. Probab. **49**, 150–166 (2012)
72. A. Lambert, Completely asymmetric Lévy processes confined in a finite interval. Ann. Inst. H. Poincaré. Probab. Stat. **36**, 251–274 (2000)
73. A. Lambert, Species abundance distributions in neutral models with immigra- tion or mutation and general lifetimes *J. Math. Biol.* **63**, 57–72 (2011)
74. A.L. Lewis, E. Mordecki, Wiener-hopf factorization for Lévy processes having positive jumps with rational transforms. J. Appl. Probab. **45**, 118–134 (2008)
75. R.L. Loeffen, On optimality of the barrier strategy in de Finetti's dividend problem for spectrally negative Lévy processes. Ann. Appl. Probab. **18**, 1669–1680 (2008)
76. R.L. Loeffen, An optimal dividends problem with a terminal value for spectrally negative Lévy processes with a completely monotone jump density. J. Appl. Probab. **46**, 85–98 (2009)
77. R.L. Loeffen, An optimal dividends problem with transaction costs for spectrally negative Lévy processes. Insur. Math. Econ. **45**, 41–48 (2009)
78. R.L. Loeffen, J-F. Renaud, De Finetti's optimal dividends problem with an affine penalty function at ruin. Insur. Math. Econ. **46**, 98–108 (2010)
79. E. Mordecki, The distribution of the maximum of a Lévy process with positive jumps of phase-type. Theor. Stochast. Proc. **8**, 309–316 (2002)
80. P. Patie, Exponential functional of a new family of Lévy processes and self-similar continuous state branching processes with immigration. Bull. Sci. Math. **133**, 355–382 (2009)
81. M.R. Pistorius, On doubly reflected completely asymmetric Lévy processes. Stochast. Proc. Appl. **107**, 131–143 (2003)
82. M.R. Pistorius, On exit and ergodicity of the spectrally one-sided Lévy process reflected at its infimum. J. Theor. Probab. **17**, 183–220 (2004)
83. M.R. Pistorius, in *A Potential-Theoretical Review of Some Exit Problems of Spectrally Negative Lévy Processes.* Séminaire de Probabilités, vol. XXXVIII. Lecture Notes in Math., 1857 (Springer, Berlin, 2005), pp. 30–41
84. M. Pistorius, On maxima and ladder processes for a dense class of Lévy process. J. Appl. Probab. **43**, 208–220 (2006)
85. J-F. Renaud, X. Zhou, Moments of the expected present value of total dividends until ruin in a Levy risk model. J. Appl. Probab. **44**, 420–427 (2007)
86. D. Revuz, M. Yor, *Continuous Martingales and Brownian Motion*, 3rd edn. (Springer, Berlin, 1999)
87. L.C.G. Rogers, Wiener-Hopf factorisation of diffusions and Lévy processes. Proc. Lond. Math. Soc. **47**, 177–191 (1983)
88. L.C.G. Rogers, The two-sided exit problem for spectrally positive Lévy processes. Adv. Appl. Probab. **22**, 486–487 (1990)
89. L.C.G. Rogers, Evaluating first-passage probabilities for spectrally one-sided Lévy processes. J. Appl. Probab. **37**, 1173–1180 (2000)
90. J. Rosiński, Tempering stable processes. Stochast. Proc. Appl. **117**(6), 677–707 (2007)
91. K. Sato, *Lévy Processes and Infinitely Divisible Distributions* (Cambridge University Press, London, 1999)
92. R.L. Schilling, R. Song, Z. Vondraček, in *Bernstein Functions. Theory and Applications.* de Gruyter Studies in Mathematics, vol. 37 (2010)
93. L. Shepp, A.N. Shiryaev, The Russian option: Reduced regret. Ann. Appl. Probab. **3**, 603–631 (1993)
94. L. Shepp, A.N. Shiryaev, A new look at the pricing of the Russian option. Theor. Probab. Appl. **39**, 103–120 (1994)
95. R. Song, Z. Vondraček, in *Potential Theory of Subordinate Brownian Motion*, ed. by P. Graczyk, A. Stos. Potential Analysis of Stable Processes and its Extensions, Springer Lecture Notes in Mathematics 1980 (Springer, Berlin, 2009), pp. 87–176

96. R. Song, Z. Vondraček, Some remarks on special subordinators. Rocky Mt. J. Math. **40**, 321–337 (2010)
97. H. Stehfest, Algorithm 368: Numerical inversion of Laplace transforms. Comm. ACM **13**, 47–49 (1970)
98. V.N. Suprun, Problem of destruction and resolvent of terminating processes with independent increments. Ukranian Math. J. **28**, 39–45 (1976)
99. B.A. Surya, Evaluating scale functions of spectrally negative Lévy processes. J. Appl. Probab. **45**, 135–149 (2008)
100. L. Takács, *Combinatorial Methods in the Theory of Stochastic Processes* (Wiley, New York, 1967)
101. A. Talbot, The accurate numerical inversion of Laplace transforms. IMA J. Appl. Math. **23**, 97–120 (1979)
102. O. Thorin, On the infinite divisibility of the lognormal distribution. Scand. Actuarial J. **3**, 121–148 (1977)
103. M. Veillette, M. Taqqu, Numerical computation of first-passage times of increasing Lévy processes. Meth. Comput. Appl. Probab. **12**, 695–729 (2010)
104. V. Vigon, Votre Lévy rampe-t-il? J. Lond. Math. Soc. **65**, 243–256 (2002)
105. D.V. Widder, *The Laplace Transform* (Princeton University Press, Princeton, 1941)
106. M. Winkel, Right inverses of non-symmetric Lévy processes. Ann. Probab. **30**, 382–415 (2002)
107. V.M. Zolotarev, The first passage time of a level and the behaviour at infinity for a class of processes with independent increments. Theor. Probab. Appl. **9**, 653–661 (1964)

LECTURE NOTES IN MATHEMATICS

 Springer

Edited by J.-M. Morel, B. Teissier; P.K. Maini

Editorial Policy (for the publication of monographs)

1. Lecture Notes aim to report new developments in all areas of mathematics and their applications - quickly, informally and at a high level. Mathematical texts analysing new developments in modelling and numerical simulation are welcome.

   Monograph manuscripts should be reasonably self-contained and rounded off. Thus they may, and often will, present not only results of the author but also related work by other people. They may be based on specialised lecture courses. Furthermore, the manuscripts should provide sufficient motivation, examples and applications. This clearly distinguishes Lecture Notes from journal articles or technical reports which normally are very concise. Articles intended for a journal but too long to be accepted by most journals, usually do not have this "lecture notes" character. For similar reasons it is unusual for doctoral theses to be accepted for the Lecture Notes series, though habilitation theses may be appropriate.

2. Manuscripts should be submitted either online at www.editorialmanager.com/lnm to Springer's mathematics editorial in Heidelberg, or to one of the series editors. In general, manuscripts will be sent out to 2 external referees for evaluation. If a decision cannot yet be reached on the basis of the first 2 reports, further referees may be contacted: The author will be informed of this. A final decision to publish can be made only on the basis of the complete manuscript, however a refereeing process leading to a preliminary decision can be based on a pre-final or incomplete manuscript. The strict minimum amount of material that will be considered should include a detailed outline describing the planned contents of each chapter, a bibliography and several sample chapters.

   Authors should be aware that incomplete or insufficiently close to final manuscripts almost always result in longer refereeing times and nevertheless unclear referees' recommendations, making further refereeing of a final draft necessary.

   Authors should also be aware that parallel submission of their manuscript to another publisher while under consideration for LNM will in general lead to immediate rejection.

3. Manuscripts should in general be submitted in English. Final manuscripts should contain at least 100 pages of mathematical text and should always include

   - a table of contents;
   - an informative introduction, with adequate motivation and perhaps some historical remarks: it should be accessible to a reader not intimately familiar with the topic treated;
   - a subject index: as a rule this is genuinely helpful for the reader.

   For evaluation purposes, manuscripts may be submitted in print or electronic form (print form is still preferred by most referees), in the latter case preferably as pdf- or zipped psfiles. Lecture Notes volumes are, as a rule, printed digitally from the authors' files. To ensure best results, authors are asked to use the LaTeX2e style files available from Springer's web-server at:

   ftp://ftp.springer.de/pub/tex/latex/svmonot1/ (for monographs) and
   ftp://ftp.springer.de/pub/tex/latex/svmultt1/ (for summer schools/tutorials).

Additional technical instructions, if necessary, are available on request from lnm@springer.com.

4. Careful preparation of the manuscripts will help keep production time short besides ensuring satisfactory appearance of the finished book in print and online. After acceptance of the manuscript authors will be asked to prepare the final LaTeX source files and also the corresponding dvi-, pdf- or zipped ps-file. The LaTeX source files are essential for producing the full-text online version of the book (see http://www.springerlink.com/openurl.asp?genre=journal&issn=0075-8434 for the existing online volumes of LNM). The actual production of a Lecture Notes volume takes approximately 12 weeks.

5. Authors receive a total of 50 free copies of their volume, but no royalties. They are entitled to a discount of 33.3 % on the price of Springer books purchased for their personal use, if ordering directly from Springer.

6. Commitment to publish is made by letter of intent rather than by signing a formal contract. Springer-Verlag secures the copyright for each volume. Authors are free to reuse material contained in their LNM volumes in later publications: a brief written (or e-mail) request for formal permission is sufficient.

Addresses:
Professor J.-M. Morel, CMLA,
École Normale Supérieure de Cachan,
61 Avenue du Président Wilson, 94235 Cachan Cedex, France
E-mail: morel@cmla.ens-cachan.fr

Professor B. Teissier, Institut Mathématique de Jussieu,
UMR 7586 du CNRS, Équipe "Géométrie et Dynamique",
175 rue du Chevaleret
75013 Paris, France
E-mail: teissier@math.jussieu.fr

For the "Mathematical Biosciences Subseries" of LNM:

Professor P. K. Maini, Center for Mathematical Biology,
Mathematical Institute, 24-29 St Giles,
Oxford OX1 3LP, UK
E-mail : maini@maths.ox.ac.uk

Springer, Mathematics Editorial, Tiergartenstr. 17,
69121 Heidelberg, Germany,
Tel.: +49 (6221) 4876-8259

Fax: +49 (6221) 4876-8259
E-mail: lnm@springer.com